FIRST EDITION

PUBLISHERS' WEEKLY SAYS:

"This is a book that can save lives; it should be read and heeded in every household in the country. Howard Owen, who is Fire Commissioner in Baltimore, thinks our fire casualty figures are too high; here he offers sound advice on how to prepare for a fire emergency. While he recommends heat and smoke detectors, he insists that each household must have an escape plan, with alternate routes. Once the family is out, they have a previously agreed-upon meeting place. Then call the fire department—from another house, or box. Forget pets, valuables; the object is to save human life. Owen discusses firefighting, fire hazards and fires in highrise buildings. For the last, procedures may be different—residents or workers should learn what to do, where to go, and follow directions. Owen may frighten us but he is convincing. Tomorrow could be too late."

I sincerely hope that FIRE AND YOU will better prepare you and your family for that inevitable fire.

Howard R. Owen

FIRE and YOU

FIRE and YOU

by Howard R. Owen

Doubleday & Company, Inc.
Garden City, New York
1977

ISBN 0-385-12372-8
Library of Congress Catalog Card Number 76-28934

Printed in the United States of America

TO THE FIRE FIGHTERS AND FIRE OFFICERS
WHO DEVOTE THEIR LIVES
TO PROTECTING YOURS

Acknowledgments

In undertaking a subject as serious as fire, I spent three years before even putting the pen to paper, and then another two years writing, reviewing, and realizing that never before had another approached the fear of fire with a touch of common sense. My experience with the men of the Baltimore City Fire Department has been most revealing.

Jack Powers has been a refreshing critic. The creativeness and research on the part of Jack Scott has been a significant contribution. Joe Sheppard's interpretive illustrations on fire are greatly appreciated.

I want to especially acknowledge the National Fire Protection Association for its significant contribution on behalf of fire prevention and specifically for its permission to quote from its bulletins and pamphlets on fire safety.

The National Commission on Fire Protection and Control through its report *America Burning* has provided a great deal of insight into the dilemmas of fire.

H.R.O.

Contents

Chapter 1	It Could Happen to You	1
Chapter 2	Why You?	7
Chapter 3	How It Gets You	13
Chapter 4	It's in Your Hands	18
Chapter 5	Decision by Indecision Can Be Fatal	21
Chapter 6	Fire Is . . .	30
Chapter 7	Plan Now	37
Chapter 8	Getting Help	47
Chapter 9	Bailing Out	50
Chapter 10	To Jump or Not to Jump	61
Chapter 11	Don't Jump	69
Chapter 12	Who's in the Red Car?	82
Chapter 13	The Fire Panic	86
Chapter 14	Should You Be a Hero?	93
Chapter 15	Save the People First	95
Chapter 16	Fires Do Come Back	100
Chapter 17	Children and Matches	103
Chapter 18	Smoking and Matches	111
Chapter 19	Heating and Cooking	116
Chapter 20	Electrical Hazards	122
Chapter 21	Sparks and Stuff	127
Chapter 22	Arson Is Potential Murder	134
Chapter 23	To Fight or Flee	138
Chapter 24	Using the "Right" Fire Extinguisher	141
Chapter 25	Getting Involved	147
Chapter 26	Alert at Work or Play	152
Chapter 27	Eating Out	175
Chapter 28	The Early-warning Detector	182
Chapter 29	Now It's Up to You	188
Index		197

FIRE and YOU

CHAPTER 1

It Could Happen to You

I am going to assume that you will be in a fire sooner or later, because it is dangerous to assume that you will not. I am going to assume first that your fire will be in your home during the night while you are asleep, because that is most likely. I will show you, by creating a myriad of examples, how you have been accepting everyday risks as you go about your work and play. I am going to tell you how to get out of your fire alive, if that is at all possible, even though I know that someone reading this book right now will make the age-old mistakes and die anyway, from fire or the ensuing panic.

Each year in the United States twelve thousand people like you die in fires, a figure twice the rate of the next nearest country. In the past twelve years 143,000 Americans died this way. Another 300,000 persons are badly burned every year, some so horribly they would have preferred outright death had they the choice. That's too high a figure; it is neither acceptable nor affordable.

These victims had several things in common with each other and one or two with you: They didn't expect to have a fire and they didn't expect to die in that fire. The majority of them didn't have to die in their fires; most of them died from smoke inhalation without a flame ever touching them. Most of them could have saved themselves if they had had early warnings of their fires and had known what to do with such an advantage. As it was, the fire and its smoke were their only warnings, and they came too late. Many simply died in their sleep from the deadly gases and never knew anything ever again.

With a combination of common sense and a choice of early-warning products now generally available, you can greatly increase your chances for survival in your fire.

I am not primarily interested in the fact that around eight billion dollars in property go up in flame each year. That doesn't mean anything to you when your life is at stake. That's ridiculously irrelevant when you think about the lives lost and tortured by fire . . . when you start thinking about your life and your fire. The one you think will never happen. No one ever does. But it keeps on happening.

The smoke detector is not a new gadget and is itself not revolutionary. What is revolutionary is that smoke detectors used to cost a lot and now they don't. For the very first time, they are generally affordable; anyone can own one.

I'm going to start by telling you to forget everything you were ever taught about fire. The basic information has been presented incorrectly a thousand times over. It is inaccurate, it is misleading, and it causes confusion and hesitation. The proof that it doesn't hit home, it doesn't connect, it doesn't work is that this old boring message has never changed the very statistics it is so addicted to quoting.

A deluge of pamphlets, folders, and cartoons have been printed on various aspects of fire safety, but none of them has taken the time to tell you why the do's and don'ts of fire safety relate directly to the people who have the responsibility for the lives and property of others, that's you, your family, your neighbors, and fellow employees.

Before you come face to face with fire, one of America's most dreaded causes of death and disfigurement, you should know that within the next hour, more than three hundred fires will burn in the United States, killing at least one person, injuring over thirty-five, crippling and disfiguring some of them for life.

It is time now for you to take some aggressive action. Only motor-vehicle accidents and falls rank higher than fire deaths and related injuries, yet the public has become much more conscious of the far less deadly problem of air pollution. The United States leads the world in per capita deaths and property loss from fire.

Now here on this page is where you can make a series of serious mistakes, many of which could prove to be fatal. Be perfectly honest with yourself and let's see what happens. As you are reading this page, you smell smoke; nobody has yelled fire. Regardless of where you are, imagine this is the real thing. The lights have gone out, it's becoming hard to breathe, and you are getting scared. Now shut your eyes and imagine how you are going to get out of this situation.

Well, fortunately, this time you didn't become a statistic, but the odds are (regardless of your education) that you either took an unnecessary risk, forgot to call the fire department, tried to put out the fire, or forgot to warn others of the possible danger. Imagine if this had been real heat, smoke, and panic. Would you have done the right thing?

Much of what I am going to relate to you is a clear-cut, logical procedure that, when followed as directed, has proved over and over again to be the safest protection for a family.

For now, forget everything about fire, except that it is hot, quick, aggressive, progressive, violent, deadly, and absolutely destructive of things and persons indiscriminately. I want you to forget everything about fire but those things, because now I am going to show you how to think about fire so that you need not die in it.

I'm not talking about fire prevention, nor am I talking about putting out your fire.

I have mentioned aggressive action. You may not think of yourself as pushy, or a leader, but I tell you that your life depends upon following a logical sequence of events and asserting your authority and responsibility in time of need.

It is most important to remember that the fire department cannot respond until you have taken some action to save yourself. That action may include an early-warning fire-detection system, a comprehensive fire-protection plan, or some sort of preplanned escape pattern.

I am going to tell you what you must do. Not what you should do, but what you *must* do. And then I'm going to spend the rest of this report telling you why and how you must do it.

I'm going to tell you how to save your life and give yourself the advantage of a *little* bit of time. The very first thing you must do, you must do now. And that is, you must accept your responsibility in this and get it really straight in your head. Your survival is your responsibility alone. You cannot depend on anyone or anything but yourself in this encounter with your fire. And you cannot allow another soul to dilute your responsibility to them. Your responsibility to your loved ones, and theirs to you, can be decided only *before* your fire. When your fire starts, any confusion or dilution of this responsibility can only cause hesitation. In a real fire, hesitation is possible death. If you do not contract with yourself and with your loved ones, it could easily follow that one of you could cause the death of another, by doing the "human," or the heroic, thing for that very person who dies of that misplaced interference.

If you make your plans now, you gain the likelihood of reacting properly, and all of you will most probably survive. At this point I must assume that you accept this responsibility; there is no other way.

OK. Your alarm just went off.

If you have foreseen this possibility and made a plan, your panic will be manageable and you will be in control. You will know what to do and everyone in the house will know what to do. And each will do it automatically.

There is no such thing as a tough exam if you know the answers; there is no such thing as an insoluble problem when you already have the solution. This is your fire plan; it will be simple and natural once you understand it:

1. Alert any others.
2. Get out.
3. Congregate.
4. Notify the fire department.
5. Do not go back in.

That's it. . . . I will repeat it, expand on it, drive it in, and make it stick.

Even at the point when you and your family are safely out of the house, you must remain responsible and notify the fire department yourself, at once. You *cannot* take for granted that someone else has done it for you. If the fire department gets seven calls on the same fire, nothing is lost. They are already responding to the first call. Many calls on the same fire cannot possibly do any harm. It can on the other hand assure the fire department that the alarm is not false.

Do *not* go back in. It is a fact that many persons who went back into their fires did not come out alive the second time.

At the point when the fire department arrives, you have no further responsibility except to obey the fire officer in charge. The firemen know what to do and they will do it without hesitation. But don't disappear; you are familiar with your home, they aren't and may need to ask some questions that only you can answer. The first one will be: Is everybody out?

The fire is now out of your hands. It will be put out and you will be cared for. You are safe.

Seeing all that smoke and fire in your home and knowing what it means will be heartbreaking. But you're alive and you know it because your family is hugging you, and you're all crying because you're

looking at where you were just a minute before. It is an empty crematorium and you're looking at each other, and you're all safe and not burned into permanent agony. You're alive and not dead and you're smiling at each other and you're crying and tomorrow will be time enough to survey the ashes because you made it . . . *all* of you made it.

Now on the other hand, if you don't have a fire escape plan . . .

CHAPTER 2

Why You?

By and large, antifire and fire-prevention literature is downright confusing and sometimes dangerous.

When I became President of the Board of Fire Commissioners of Baltimore, Maryland, I undertook, as my first obligation, to study everything I could about fire so that I could responsibly apply myself, in some order of priorities, to solving the problems that came to me with the office.

Being human, my natural orientation is to life. The loss of life is the most pressing and tragic problem confronting fire departments today.

I read and read. I watched television tapes, movies, and training films. I went to fires, constantly, any fire, anytime. I studied the equipment, the fire fighters, the methods, the attitudes. I questioned, I listened to everybody and absorbed all the information I could find that could possibly reveal to me how to solve the problem of saving lives. I reviewed literally hundreds of the brochures, that "official" literature generated by insurance companies, fire departments, and fire authorities over so many years.

I came to realize that those brochures are exceptionally revealing and seem to be more directly related to our perennially high fire-death toll than I had realized—in a confusing and destructive way.

1. With few exceptions, they seem to be written by the same hand . . . a word changed here, a juxtaposition there, a frequent change of the cast of stick figures.
2. They succeed in turning off the very readership to which they are directed.
3. They are generally incomplete and don't motivate people into action.

4. They are often contradictory and inconsistent.
5. They seem to have no positive effect in lowering the life loss.
6. They are basically concerned with the dollar value of the fire loss.

It is amazing to me that so much of the literature is inclined to accept the fact that some loss of life is acceptable. Quite clearly this attitude is leaking through the message and causing even the better literature to lose its impact. You can't order people to do even those things that might save their lives, and that probably is fair and reasonable. If you don't relate to people, it is unreasonable to expect them to react to your message. So you will note throughout this book that I will request, even push and shove you toward, lifesaving procedures and routines, but as I do so I will offer you a meaningful and logical explanation as to why this may save your life.

My main concern is to save as many lives from fire as possible. Early-warning detectors are the best single lifesaver in sight, even though they will be improved over the next few years. Testing, evaluation, research, and development will be accelerated and improved by the enormous increase in their acceptance and use. Despite this, you cannot afford to wait for those improvements.

It is clear to us that the victim stands much less chance for survival if he is disoriented when the initial alert is received. Whether he is asleep, taking medication, drugs, or is just plain drunk, he can't function until he realizes what is happening.

Most people who die of fire die at home—in their houses, apartments, rooming houses, mobile homes, house boats, whatever is home.

They mostly die at night while they're sleeping in their bedrooms, wherever they sleep. They die coming from their beds trying to reach safety.

How do they wake?

They wake up coughing in the darkness, that bad dream of smoke and fire still clinging, sticking to them . . . then dive back deeply into sleep to try and shake it off.

They sit straight up in bed, paralyzed with recognition, jaws too locked in horror to cry for help, mouth too dry to shout a warning, mind too frozen to think through motions of escape, arms too rigid to touch and try to save the sleeping form beside them, sleeping on. . . .

J.S.

They leap from bed to run, fall over, slapped down by heated smoke already settling thick and deadly just above the bed. They stand transfixed in terror before their only door which is leaking poisonous snakes of smoke and flame. They cry and shout and call their loved ones, who do not answer, or cannot. They wrench the door wide open letting the superheated air, the flames, the poison smoke into the room, into their lungs as well, and fall, not even knowing when they hit the burning floor.

This report is about choices . . . choices not nearly so limited as the nightmares recounted above. The more time there is the more choices you have. You do not have to wait for your fire to announce itself by knocking at your bedroom door.

If they wake up at all, how do people first know of the fire? They smell smoke. They hear the fire; if there is a strong fire going near enough or big enough, they can hear crackling, popping sounds. They can hear glass breaking. There may be an explosion: gasoline, a propane tank, aerosol bombs, some exotic fuel going off. They may hear timbers collapsing. They may hear outside noises: neighbors hollering, animals barking unusually, running around loudly, bumping into things, howling, clawing at some door. Someone else might be coughing involuntarily from the smoke. They may hear signaling, calling, screaming. Someone may be yelling "Fire!" The arrival of the fire trucks is often the first sound they hear. Hardly an early warning!

In a curious way, when I became commissioner I had an apparent advantage I might not have had if I had come up through the ranks as a fire fighter. Suddenly, I was on the inside with 2,300 fire fighters and $20 million worth of fire-fighting and rescue equipment in an area with 962,000 residents and 203,000 dwelling units. I had to know how to maximize the efficiency of our fire and rescue service to reduce our vulnerability to fire death to the absolute minimum, whatever that is.

It's like the general budget syndrome: If any agency is assigned a budget, it will expend that budget. Well, we as a people are being told to accept a death budget. Now, don't get me wrong. Being human and alive, we are mortal and must die. Each of us *will* die someday. I insist we have some choice in how we will not die. Fire death and death from fire injury are among the most painful and horrible ways to go. I am concerned with life and death, as related to fire, and I'm saying that it is possible to change the fire-death and fire-injury figures without lying with statistics.

I came to this job as a problem solver and I learned about fire. Not everything, but certainly enough to know that saving life from fire is our biggest problem. I came to be absolutely convinced that it can be solved. Not that nobody will die from fire or smoke. There are always accidents: explosions, circumstances of death and injury that cannot at this time be humanly preventable or foreseeable.

This is not a report about the fire department . . . especially this is not a book about the Baltimore City Fire Department. I happened to have gained my experience about fire in that city. But it does not matter whether you live in the country, on a mountaintop, in a town, or on a houseboat. It does not matter whether you are served by a volunteer or paid fire department, a bucket brigade, or a helicopter fire and rescue service. Any fire department is more effective than no fire department. They are all good. They're the best you have and they can do their job better than you can.

I am going to tell you about fire and rescue departments, but only to clear up some of the mysteries and superstitions that are generally prevalent. And I am going to do that only as a further description of your responsibility to yourself; and I must make it clear that the responsibility of the fire department is only secondary so that you won't die from an unrealistic dependence on them.

With the wide availability of early-warning detectors, more than half of those whose lives will be in jeopardy, including yours, could be saved. But *no one and nothing,* not even the smoke detector alone, can save you. Each must save himself!

No wonder Americans are described as apathetic and indifferent to their fire problems. On the one hand, they are exposed to movies and television depicting and encouraging heroism and superhuman courage as the normal and desirable approach, and, on the other, they are furnished with pamphlets from insurance agents, school, plant fire wardens, Sunday schools, even from fire departments, that indicate that fire deaths are inevitable. This literature confuses and insults all who read it. It certainly does not increase your ability to understand *why* you must undertake the responsibilities now. You are probably better off in relying on your natural instincts of survival and your own common sense and intelligence.

This report is about death prevention and burn prevention in your fire *situation.* I insist on assuming that your fire will happen and that it will probably be while you and your family are asleep. It will happen when it's too late to clean up the trash, put away the gasoline, check the wiring and all that. Your fire has just erupted and you are asleep.

Further on we'll discuss your day-alert fire at homc, at work, in the garage, or wherever. But a daytime fire is less dangerous because it will happen while you are awake and alert and your behavior then will be different.

Before we move on I want to tell you I think fire is fascinating. A chunky oak log warming one's soul and toes on a cold winter night makes the world a better place. A cozy campfire puts you close to something elemental in human nature. Most people feel that. That's probably why children are drawn to fire initially as a wonderful and irresistible thing. That is why both faces of fire must be carefully revealed to the youngster. If the child is scarred by a bad experience with fire, he or she will lose that value of the beauty, the utility, and the profound fascination fire has held for humans since Prometheus lent it to us.

CHAPTER 3

How It Gets You

I think the next logical step is to explain how fire kills. Heat rises when a fire starts in the basement or ground floor and if you have an open stairway, smoke, heat, and gases will travel that route. If you are in a second-floor bedroom, for example, with your door open, the smoke, heat, and gas will enter and rise to the ceiling and work their way down. Breathing this high concentration of gases can cause your death.

The flames of fire are not the leading killer, and the recognition of this factor should be seriously considered when you set your priorities for protecting yourself and your family.

Fire requires oxygen to burn; so do we as human beings need a concentration of about 17 per cent to function in a rational manner. Fatigue, loss of muscular co-ordination, and ineffective activity become prevalent when fire causes the oxygen content to fall below 16 per cent. As the oxygen content continues to decrease, breathing will cease and death will come within ten minutes. This is known as asphyxiation and is the major cause of fire deaths.

Unless you have seen it happen it is not possible to believe how fast the temperatures can accelerate from even a small fire; 300° F. is sufficient to cause loss of consciousness or death within several minutes. Superheated air and gases, combined with a high moisture content, destroy or seriously damage lung tissue.

Many times, the effects of asphyxiation and superheating happen so fast that unless you are protected by an automatic early-warning device, you may never even have a chance of escape or rescue.

Smoke poses a double problem: It moves fast, blocking the visibility of exits; and the inhalation of thick smoke, irritating the mucous membranes of the respiratory tract, is a serious situation, often fatal.

Toxic gases are created by both the fire itself and by the incomplete combustion of the materials burning. The various combinations of these poisonous gases, in a state of intense heat, may cause breathing to be impossible. Quite likely, these elements will create a blinding effect and cause the victim to become rapidly disoriented.

The public has been fooled over and over again by the flames of fire. Don't be lulled into a false sense of security because you have located the fire and its flame seems to be of manageable size. By this time, heat is building up and any of the other lethal side effects may be playing their parts. Remember that all of the other major killers can get you before the flames, even if they never become visible.

People do not understand the intense heat that can be created even in a relatively small fire, especially if the heat is confined within a given area. A telephone can be melted to an unrecognizable glob without real fire in the room, just by the intense heat. I recall a fire where an elderly woman died. The inlaid metal in the furniture, good metal, actually melted right out of the furniture. Heat of a magnitude that will melt such plastic and metal obviously will kill in a matter of seconds.

When a person awakens amid smoke, he tends to rush to the door without any thought of what may be on the other side of it. He grabs the door, pulls it open, and one deep breath of the superheated air and gas from the other side is enough to be fatal. The typical bedroom door is not fire-rated (i.e., not so constructed as to be rated as fire retardant). A closed door performs two functions: It actually stops the fire from traveling from one side to the other and it eliminates the draft, which the fire will follow. Fire will follow a draft, whether the draft is caused by an open stairwell, an open window, an exhaust fan. If there is any indication of heat, fire, or smoke, approach the door with caution. Do not just grab the door and yank it open. Watch for smoke around the edges of the door if you are able to see. If it is dark, feel the door, the knob, even the wood, and then open the door only if you feel no heat. Even then, open it carefully. Be prepared to slam it shut again if the heat and smoke are there and you can't get through by that exit. It is very important that you brace yourself before opening the door, in case pressure from the other side makes it difficult to slam the door quickly.

Is it conceivable that it can be two hundred degrees hotter on one side of the door than on the other? Yes. An ordinary door will continue to hold out heat and smoke for a while. Once you open it though, if you don't shut it quickly or get out quickly, you may be

J.S.

trapped. I have seen instances where even very flimsy bedroom doors have kept out the fire until the firemen got there. If there had been people in that bedroom, they would have been saved. I recall another fire where every room in the house was black, charred, and gutted, except for the rear bedroom, where the door happened to be closed. When you walked up the stairway, the door was almost completely charred from the outside, but once you opened that door and went inside, you would never know there was a fire in the house. That room contained the only survivors. So remember that running down a hall filled with fire and smoke could make you a dead hero instead of a slightly scarred victim who waited behind a closed door.

Because a door can offer some protection, one should sleep with his bedroom door closed. Of course, that runs contrary to a mother who wants to hear where the children are at all times, who wants to know what is going on in the house, but a door can be a definite fire stop. We know that a closed door will at least for some time stop the fire. As many times as I have talked to civilians and explained this to them, the very first question they ask you is "How will I hear my child if the door is closed?" It is a very hard question to answer. Even though you explain to them all of the things that will happen if the door is open, the first thing that comes to their minds is: "If I can't hear my child, how can I help my child?" You can't beat that parental instinct to get to the child to save it. But if you have taken the time to create a comprehensive preplan for the household, you will probably be able to do the right thing.

Another condition prevalent in the large majority of fires is the pressure built up by heat and smoke. We have discussed the reason for bracing the door when you open it. If the door isn't braced, there is a possibility that pressure can push the door wide open and you will have intense smoke and heat right in your face. Sometimes a fire begins to build but doesn't have sufficient oxygen in a closed area to really get roaring; then when you open the door or break that window, you let in enough new oxygen to cause an outright explosion.

A fire that cannot get sufficient oxygen to really flame can generate large amounts of carbon monoxide gases. Gases and the heat can build to very high levels without a rip-roaring flame. This is why fire fighters, when they arrive on the scene, will, if possible, ventilate from the top. That is to let the heat, smoke, and gases out.

Heat and smoke both travel in the same direction; they rise to the highest possible level from the origin of the fire. Once they reach the ceiling or whatever the uppermost level is, then they start to build back down toward the floor. This is why you get as close to the floor as you can. That is where you are going to find the best possible air until you can get out.

Obviously if dense smoke is present, your best vision also would be at a level as low as possible. It is quite conceivable that, by the time you became aware of the fire, the smoke was already down to three feet above the floor (which is the bed level) and you are in trouble if you try to stand up. There could be a tremendous difference in temperature between the ceiling and the floor and that difference can be as much as 300° F.

Quite obviously there are a multitude of situations that rapidly develop in a fire emergency, many of which can cause very serious injuries or death. If you are going to have a fair chance of escape, you must have a well-devised and well-practiced fire plan. Somebody has to take the responsibility before the unexpected strikes your life. Those around you will depend upon your taking positive action.

CHAPTER 4

It's in Your Hands

The head of the household has the responsibility for being sure that there is a fire evacuation plan that has not only been drawn up but has been accompanied by fire drills. He or she must be sure that everybody old enough to fend for himself understands what procedure he must take at any time of the day or night.

I think it should be clearly stated that the only responsibility you should be concerned with is for the life and safety of the occupants of the household. Forget the pets, forget the television set, and everything else. You have only one obligation, that is to get out and summon help. Always remember that the correct order is to get out first, summon help second.

There should be at least two routes of escape from any given location within the household. The obvious escape routes are the normal hallways and doors if they are accessible. If they are not accessible, some kind of alternate route or combination of alternate routes must be mapped out.

Before any plan can be put into action, the initial warning must be sounded. Such automatic devices as smoke and heat detectors are now readily available. Before you have intense heat in a household, you are most likely to have smoke. The heat detector is not as likely to give as early a warning as the smoke detector. By the time the fire has advanced enough to create sufficient temperatures to trigger the heat detector, the fire may be out of control.

The other early warnings discussed earlier may occur: crackling noises, breaking glass, the smell of smoke, outside noises, barking dogs. The investigation of one of these warnings has saved many a family from disaster.

Coughing by someone else in the household may indicate that that person is already being overcome by smoke. There is one other last warning: the arrival of the fire engines. The thing to do is to start your evacuation plan. Don't wait for a confirmation by alarms or other sounds; delays lessen your chance of survival.

Don't forget, in your haste to escape, after discovering or being alerted to a fire situation, that a warning should be signaled to the other occupants. Don't shout "Fire" unless you mean it, but the quickest possible word that everybody understands is "Fire," not "I smell smoke," "Help," or "Get out." If you think that there is fire, say it regardless of where you are and say it loud and say it clear. Don't be indecisive.

You should always discourage the shouting or yelling of "Fire" as a game or prank by youngsters. Imagine the panic that could be caused in a crowded movie house. Because of the seriousness of the fire alarm, many instances occur where people send false alarms causing lives to be risked where no useful purpose can be served.

The Baltimore City Fire Department had a case where a gentleman, locked from his apartment, asked his neighbors to call the fire department. The neighbors called the fire department and said that there is a man at the window shouting "Fire." The fire fighters responded. The man was taken in for a false alarm and for giving a false fire report.

In many cases, women in metropolitan areas are being encouraged to yell "Fire" if they are attacked on the streets, realizing that a bystander is more likely to respond to a fire plea than a cry of rape. If the cry is "Fire," the first impulse, however, is usually to alert the fire department, which really can't help in this situation. "Fire" is a clear simple word understood by all, from the very young to the elderly. It should be used only as the universal alert for a fire. The death rate from fire among children under five and the elderly over sixty-five is three times that of the rest of the population. Though together these young and old make up only 20 per cent of the American population, they account for 45 per cent of the fire deaths. Anything that makes the early-warning alert clearer for these groups will make a significant reduction in loss of life and serious injury.

Assuming now that you have received one or more early warnings, the first priority is to get out of the house and be sure that everybody else is out. This will come about by an evacuation plan that has been practiced over and over again in the form of fire drills. It has to be clearly understood that the responsibility first is not to fight the fire, but to get out of the house and be sure everybody is accounted for.

Alerting the other occupants is a must, after the initial finding of the fire situation. Anything that can make noise to arouse the other occupants will do: a police whistle, a dishpan with a large spoon, or banging on the wall. Some kind of a prearranged tapping noise might be one way. It is extremely important that everybody in the house knows what the signal is going to be, and it must not be used for anything else. Don't count on your voice as the only means of signaling. It is possible that the smoke and heat have affected your speech.

If you wake up and become aware of smoke, the normal reaction is to first go through the house to see where it is coming from. Alert the others in the house first. This is imperative; over and over again we have seen the "household detective" cause a disaster by investigating instead of alerting.

CHAPTER 5

Decision by Indecision Can Be Fatal

Let's take a typical home situation with a family of, say five, including two adults, a couple of children, and let's say they have a grandparent or someone like that living in the house, not an invalid. What would be the procedure when a fire breaks out in the house during sleeping time, which would be roughly between 9 P.M. to 7 A.M.? What should a family plan for, what do they do in an evacuation plan, in what chronological order should they respond? The first thing in any plan usually is to make sure everybody is together, to get everybody out of the house (only then call the fire department). *Do not* go back into the house. The evacuation plan should not just be casually discussed in the household. You must preplan and practice the evacuation routine. The whole family should know how the plan works and take part in the fire drills so that everyone is thoroughly familiar with the plan. This is not a function that will be accomplished by sitting around the dinner table one night discussing it; it requires fire drills and a logical sequence of events that everybody in the family understands.

You must actually put it into practice, to see if it really works and to really imprint it deeply in your minds. For real, not make-believe . . . It's not a game.

You have to come to the realization that regardless of the intelligence of a person, fire is going to be a foreign thing to him; he probably has never been confronted with smoke or fire before. If he hasn't made plans, the chances of his making an error in decision-making are much greater. If he has thought the thing out and has trained himself and his family in what to do, they are much less likely to make an error.

Many times when fire fighters are confronted with an unprepared house, a dwelling that has not had prefire planning, they find children hiding or trapped under their beds. The victims have done basically the wrong thing in their panic, and it causes situations where the firemen can't execute a recovery or make use of their lifesaving techniques.

Without any prefire planning at all, most people, especially children and the elderly, will go to the door. When they smell smoke, the first thing they do is head for the exit they are most familiar with, because they never have given thought to another exit. If they get to that entrance and it is blocked, they don't know where to go. They next go to the nearest window, where fire fighters find children and adults crouched by the window sills inside a burning building, doing nothing to help themselves. Most fire fighters look inside first and ventilate as quickly as possible, expecting to find people under the window sill. The panic sequence is to go to the door first, then to the window, and then to hide themselves under beds, cupboards, or even wrap up in blankets. They have been found in shower rooms. They get confused and they hope any door will offer escape. In a panic, children often retreat to their bedroom to hide in a closet where they think they are safe from fire and smoke. Many a child has been lost in this fashion.

A few years ago, we gave lectures on fire prevention to the youngest of children. As part of the presentation, we had a situation skit in which the children were asked to respond to an imaginary fire. Later, at the fire school, I put a similar situation to the engineering group. I got a response that was outright scary. The situation I put to the engineers was as follows: Imagine you are a second-grade child coming home from school. You are entering your house. You open the door. You smell smoke. You go to the foot of the stairs, to the basement, open the door, and a big cloud of smoke comes out. What do you do?

I would call the fire department. Where? How? We got all kinds of answers. Go down and pull the box. I said, How about pulling the door closed first? Oh, yes, sure, yes.

Then what would you do? Then I would call the fire department. How about letting somebody in the house know that there is a fire? Oh, yes, sure.

You see, the sequence of events is so important, but these guys, educated as they were, were no more prepared than the second-graders. It is incredible that an engineer with all that training could not think out the situation any more clearly than a six- or seven-year-old in the second grade.

I trust by now that I have convinced you that since it is your fire, it is your responsibility to escape it. The fire department can't fight it until they know about it. As quick and as intensely committed as they are to respond, you might be dead before someone else notices your fire and calls it in. They're responsible for saving you and putting the fire out, but they can't do it until they get there. And you might all be dead by the time they get there. That's why the death rate is so high. That's why it is your responsibility. Because there is nobody else who can take it. You have to think. You have to prepare a plan, a real plan, involving you and your family and your real situation. Your fire will be real. Your plan to escape from it must be realistic or it won't get all of you out. You have to think ahead and really imagine that fire and how many different ways it can happen. You will have that many different ways to plot your best escapes. You will have to think hard about what you have to do to survive by the ways that seem best, and you must practice them until they are automatic and foolproof. You will have to work the errors out of the plan and then practice it until it becomes second nature. You're fighting lots of out-of-date instruction and much misinformation. Habits are hard to break.

Remember, when you wake up and your house is on fire, you face panic, fear, confusion, and disorientation. Your mind is going to leap to the most comfortable solution or the easiest one, and it may be a childish one. You have to fight that and win; you can't hide under a bed or in the closet. Your chances are much better if you respond directly and without hesitation. Your escape response must become as by rote, and automatic and immediate. You must train yourself to do this and everyone in your family must do the same. Because of the panic, the anxiety for yourself and your family will give you no time to put it all together later. Your error will be higher and you will be too dependent on luck. Luck can go in any direction, and the only way you want to go is out.

You have already taken the next logical step or started to. I will explain how fire kills, how smoke kills, how carbon monoxide and other deadly gases kill, how heat kills . . . even how shock kills. I will explain how fire burns human flesh as readily as any other object in the house, and I will relate all these elements to your home. Most people regard fire as abstract, as having no reality touching them beyond campfires, cooking, fireplaces, and cigarette lighters.

When you awaken and find smoke, heat, and flames, your first tendency is to rush to the door, without any thought of what may be on the other side of that door, to grab the door, pull it open . . . and burn to death. Right on the spot. Oh, it takes a little while for your body to char or roast, or bake or fry . . . but that first deep, involuntary gasp of superheated air is enough to collapse your lungs instantly, to glue them together like collapsed balloons, irreversibly. You might be conscious for a moment, but you can never unstick your scorched lungs.

Let's talk about doors for a moment. There are all kinds: fire-rated, nonfire-rated, loose, tight, thick, thin, hollow, solid, paneled, flush. What they all have in common is that any door is better than no door. Any door closed is better than any door open. An open door is the same as no door. That's why you should always sleep with your door closed, no matter what kind of door you have to close.

"My wife would never hear of that. The children . . ." This is one of the most emotional reactions to any of the rules of survival; it's a tough decision. But any door is a fire stop. It gives you time. It gives your children time. It can mean life in a death-dealing fire. Remember that it's both the parents' and the children's lives in question. Remember that it's downright inconsiderate for those kids to have to learn that Mommy and Daddy didn't make it out of a fire.

Let's get back to that bedroom door. You wake up and smell smoke. If it's bad, roll out of bed and stay low. The worst smoke and gases rise; as they thicken they then settle downward closer and closer to the floor.

Go to the door if that's in your fire plan; it's usually your best way out. Test it. Don't open it. Test it. Look for smoke around the edges if it's not too dark. The cracks, the keyhole, particularly the top and bottom. Feel the door, both the knob and the wood.

If there is significant smoke pouring in or if any part of the door is hot, brace yourself against the door. Use your weight, your mass, as a doorstop in case there is hell on the other side.

Listen. Sometimes you can hear fire.

Use your foot, your hip, your shoulder; dragging furniture will only waste time, energy, and breath. Wedge your body against that door and slowly, cautiously turn the latch and ease the door open as if there were a cobra or a tiger behind it.

If the heat or the smoke tell you very clearly that you shouldn't go out, firmly close that door and double-check the latch. Remember, fire and gas pressure from the other side can knock you on your ass and end your life in a second.

I recall an apartment house fire that was like a blow torch. The fire shot out the doors along one side of the hallway. Yet the doors across the hallway, only eight feet away, held. Those doors held and there were fifty or sixty people on that side who were kept safe just by keeping their doors closed.

When you're talking about doors you're talking about two things actually. You're talking about stopping the fire from traveling from one side to the other. And you're talking about eliminating the draft, the natural flue the fire will make and that the fire will follow.

It depends on how tightly the doors fit and some other things, but, basically, the fire will follow the draft, the door will block the draft, and the fire might well choose another way up and out.

Fireplaces exist because they allow the fire to go the way it wants to go, up. If you attempt to force it otherwise, all sorts of unpredictables occur. They are all according to natural law, but, if you are pressed for time, you might not come up with a quick answer that is valid. That draft, that flue is going to be up an open stairwell, a vertical opening; fire can be sucked horizontally along an attic or a floor from one side to another by the pull of an exhaust fan.

An explosion can be triggered by opening that door. As a fire begins to build, if it doesn't have sufficient oxygen to get roaring, it becomes a smoldering mass generating superheat and an unbelievable amount of smoke.

If your windows are open when you open that door, say you open it all at once and not just a crack, it is entirely possible to feed that fire just the right amount of oxygen to trigger a so-called smoke explosion caused by the sudden back draft. When you let the fresh amount of oxygen into the fire suddenly, you can have an outright explosion where the flames literally explode toward the oxygen source.

If that door is hot and smoke is really pouring in and that is your first planned exit, then you can thank God you've worked out another exit, because *you must not open that door.*

Your second exit may be the window if it's close enough to the ground or you have preplanned this route of escape. There is always a chance of an explosion or the door blowing in when you open a window. It's possible, but you have to look at it this way: Your first line of escape is completely blocked by the fire. If the window is your second prechosen exit, the real question is, Do you have a third option?

You could stay in the room near the windows and wait for the fire department to rescue you if you were sure they would get there in time and be able to find you before it is too late. Your exit should be no more dramatic than necessary, but it should be by your choice. If you have had early enough warning, you will test the window cautiously, opening it first a crack, then gradually. Yes, you could have an explosion but you will probably die if you stay there, so generally speaking I'd say use the window if you really have no other choice.

In a fire, even the safest way out can be dangerous; you must always choose the least dangerous exit. You don't want to kill yourself saving your life. Be careful. The first thing that comes to mind is whether you can get to a roof of any kind—above or below you—to get yourself farther from the fire.

It's probably a long shot, but even a tree or a pipe on the side of your building might be better than nothing. You will have to judge that. You'd probably be better jumping than falling; you have more control over how and where you land.

Another thing that comes to mind very quickly is the safety ladder. Unless it's provided in your prefire planning, it's unlikely you're going to dig one up during the fire, but it can be a way out. There are a number of possible flaws to the safety ladder and attempts to escape by such means can be dangerous. These are discussed at length in Chapter 10.

I haven't mentioned fire extinguishers and don't really want to get into that subject until later, but it is possible to fight your way through the fire with an extinguisher, thereby creating a safe exit route. If you have to choose in your prefire planning between affording a fire extinguisher or a smoke detector, the smoke detector is your wisest choice. It gives you time, perhaps even enough time to make a mistake and still escape. To make a further comparison: If I had a choice between a smoke detector with an escape plan, or fire insurance, there's no question I'd do without the insurance. That money is no good to me dead. And in your case, too, I'm sure your family would rather have you than the money.

Let's go back to that room. The door is too hot and smoke is coming in. What's in that smoke can easily kill you. Carbon monoxide

(CO) is the result of incomplete combustion. If fire is deprived of enough oxygen to burn fully, it will generate a lot of carbon monoxide in the smoke as well as other poisonous gases. CO is invisible and has no odor. But you better believe that if you can smell smoke, you are breathing a lot of carbon monoxide. That means that you, like the fire, are not taking in as much oxygen as you should.

Deprived of sufficient oxygen, your brain quickly loses its ability to think clearly. You will first become drowsy, fatigued, and confused. The more CO you breathe in, the longer you breathe it, the less energy and clarity you will have to escape. Your vision will blur, your judgment will get worse, and finally you will not even be able to move.

Smoke is bad enough. You can be in a smoky room and not be able to see a street light out of your window because the smoke is so thick. Carbon monoxide is different. The smoke itself can be very thin, but the CO content, invisible, can produce the same visual effect as heavy smoke. Because of the buildup of it in your blood and brain, you might not be able to see that street light through your window or even see the window.

Most home fire victims die from smoke and gases, by asphyxiation. When the firemen arrive, the fire may be consuming the victims' bodies but they died of asphyxiation.

Any fire death is horrible. And some of the burns even worse. That's why when firemen arrive the first thing they do is look for survivors in the most logical places: under the windows, and around the beds, doorways, and closets. At the same time they do a very misunderstood thing; they ventilate the fire. They immediately start tearing the hell out of the roof or some high part of the house, to create a chimney so that the fire can get directly out. In doing so this pulls air *into* the house so that survivors can breathe. The fire can then follow its natural direction, which is up, and the heat buildup in the closed house is then hopefully drawn *away* from any survivors.

Before the fire fighters go for the source of the fire they must give that fire a path over which they have some control. Roof ventilation draws the heat up so the firemen can enter the building quickly and search for victims before they actually start putting the fire out. In actual practice, all of this usually goes on pretty much at once; but the first purpose is always the rescue of human life, and *then* to put out the fire.

I hope the reason is now at least pretty apparent. We'll go deeper into that subject later. For now, look at yourself in the bedroom, up too high to get out the window. Your first exit is blocked; your second exit is blocked.

When the firemen chop that hole in the roof or somewhere above you, the pressure on your door should reverse; the heat and smoke and pressure should then go away from your door, away from you, and fresh air will be drawn *in*, through the window.

For the firemen's purpose, it does the same thing for them. They can't walk into all that heat. Ventilation draws all that heat and smoke in and up and away from them too, so they can begin the search and rescue.

We're really not the wrecking crew. We can't fight fire from the outside very well. That's how we have to get in.

Heat and smoke travel in the same direction, which is away from the fire. As the heat is carried along, the fire spreads. The heat and smoke will go to the highest possible level away from the origin of the fire, whether that level is the third-floor attic, the second story, or the first-floor ceiling. Only when it reaches that level, when it can't keep going up and out, will it then start to build back down toward the floor, any floor. So while you're escaping from the fire, you ought to stay low, as close to the floor as possible. The heat and the smoke and the gases will be the thickest closer to the ceiling.

But all of this smoke and heat will move downward toward you if you don't get out quickly. The smoke level could be down to three feet when you wake up. To be safest, when you hear the early warning, even though you would not expect the smoke to be that low, you should roll out of bed onto the floor until you are awake enough to test the situation.

If the smoke layer was down to three feet and you sat bolt upright or rushed up out of bed, you could easily take in a lungful of hot smoke that could lay you out on the floor unconscious.

By now, I am sure you have realized that it is necessary to get all this straight yourself before you can get it across to your family.

It's all logical. You must not hesitate. Fire accelerates. While you are trying to remember what to do, your chances are burning up, your choices are getting scarcer and scarcer. To summarize:

Get out. Forget the pets—the parakeet . . . the hamster . . . the gerbil . . . the dog . . . the cat.

Forget the deed, the documents, the jewelry. There is no time for it in your actual escape.

When your fire starts and your smoke detector goes off, only one thing is important in that house: your lives. Alert everyone in the house and be alerted. Then get out the best way possible.

Congregate according to plan at a specific place outside to make sure you are all out. It's too tragic when someone already out goes back in to rescue someone else who is already out and safe—and dies.

Then call your fire department.

Don't go back in.

Don't disappear. The firemen need you. You know your house better than they do. What you know about it may help them fight your fire.

They need to know if everyone is out. Otherwise they expose themselves to unnecessary danger by trying to rescue life when there is no one inside.

CHAPTER 6

Fire Is . . .

Some say only three things cause fires: men, women, and children. Not so!

In actual fact three things must come together at the same time to start a fire:

1. Fuel: paper, gasoline, bedclothes, clothing, rugs, floors, furniture, in reality much of the contents of your home environment.
2. Oxygen: normal breathing air is ample to do the job.
3. Ignition temperatures: high heat can cause a fire even without a spark or open flame.

Without these three things in combination, it is impossible for fire to occur.

But it is true that men, women, and children have a deadly habit of bringing these three fire ingredients together or at least allowing them to come together by not preventing them.

What is this thing fire? Most of what we know is merely a description of what we know it can do and what we know to count on that it will do.

In a simple log fire in the fireplace, it seems that the wood itself is burning. That is not exactly true. The vapor given off by the heated wood is doing the burning, not the log itself. This is true of nearly all things that seem to burn.

When heated, paper does not actually burn, but gives off vapor that burns even at some distance from the paper itself. Many solids, like candle wax, must be melted into a liquid before burnable vapor can escape. You heat the wick with a match; the wick vapor burns and causes further heat; the wax melts and flows up into the wick where it turns to vapor or flammable gas and there it burns well enough to keep the entire process going. Once this gets going, the flame is fed and self-sustaining until the wax is consumed.

If you snuff the candle, the wax and wick stay hot long enough to continue producing flammable vapor hot enough to be relit with a match at some distance from the wick.

Many liquids, such as heating oil, must be preheated before burning can take place. The temperature at which a fuel of any kind begins to give off vapor that can be ignited is called the flash point. The temperature at which the process will keep itself going, where the candle will stay lit, is called the fire point. Usually there are only a few degrees difference between the two.

Many materials have very high flash points and there is no way you can light them with a match. Heating oil and other household oils and materials must be heated to over 400° F. before they will burn. Many household materials, however, such as alcohol, paint and paint thinner, cleaning fluid, and gasoline (which should *never* be stored inside), have flash points so low they will light dangerously at room temperature because they are constantly giving off ignitable vapor. Gasoline is probably the worst offender: It will flash at 40° below zero. This vapor, lighter than liquid gasoline, considerably heavier than air, is so thick you can see it on a hot day, pouring out of the container from the surface of the liquid gasoline.

The flash point or fire point of the vapor is not the same as the temperature needed to ignite it. The heat source must be considerably higher than the flash point to actually ignite it, or the fuel will simply continue to give off vapor without burning, although usually at a higher rate. That is, the air temperature could be at, or above, the flash point of gasoline, but the gasoline would not ignite simply because the temperature was above —40°. Flash point means simply that if a hot enough ignition temperature is present, the vapor is willing to burn. If the fuel temperature is lower than that, it will not burn even if a very high ignition temperature is applied.

If not enough oxygen is present, regardless of temperature, fire will not occur. Under normal conditions, a flame draws enough energy to burn from the air. But in an enclosed area, like a hallway or a room with closed doors, the supply of oxygen cannot keep pace with the full potential of the fire. In such a setting the fire burns slowly, with a great deal of smoke; occasionally it will die of suffocation, much the same as a person would.

Under normal conditions, however, the oxygen of the air combines with vapor in correct proportions to sustain active fire. In open space, the ready mixture of the two, vapor and oxygen at ignition temperature, is in perfect balance and there is little smoke. In a house fire, if one increases, there is an abundance of the other available to increase the fire. And so fire once ignited accelerates, as long as there is an abundance of fuel and air.

The greater the flow of vapor, the greater the mixture with oxygen and the larger the flame. This action is caused by the heat of the flame. The hot-air currents rising from the flame create a draft suction, drawing a steady flow of oxygen into the flame area. Restrict the flow of oxygen by closing a door, for instance, and combustion becomes incomplete, retarded.

Gasoline vapor can be lit only where it mixes with oxygen in the right proportion. Between the rich and the lean mixtures there is a broad area which I will call the ripe area: the explosive range. Gasoline vapor will explode, violently. The explosive range can stretch far from where the gasoline is actually stored, to anywhere in the vicinity.

Gasoline evaporates so very rapidly, even at low temperatures, that its vapor readily overflows any crack or opening of its container. Being heavier than air, it flows downward or drifts for considerable distances on currents of air. That's why gasoline is so dangerous. It is the presence of gasoline vapor that causes a half-full tank of gasoline to explode; a nearly empty tank will explode even more violently, whereas a full tank will merely burn.

So much for fuel and oxygen for a moment. Let's get into the third ingredient: ignition temperature. With fuel at its flash point and its vapor combining readily with air, the mixture is now in a state of readiness. This is dynamite. In actual fact, one gallon of gasoline in a complete state of readiness is equivalent to sixty-six sticks of dynamite.

Yet combustion cannot occur until more heat is applied. The heat of a flame or a spark will provide the added temperature necessary to ignite the mixture. Either spark or flame provides a temperature much higher than is actually needed to ignite such a mixture in its state of readiness.

A tiny electric coil is more than enough to ignite gasoline vapor. A soldering iron, well below flame temperature, can ignite paper, paper can ignite cloth, cloth can ignite wood, and so on through the house.

Gasoline can provide an explosion hot enough to ignite everything in sight. There are some household chemicals which can explode in combination to the same effect. Unheated fuel oil cannot be ignited with a small flame, but primed with gasoline the oil vapors gradually ignite and then accelerate and the fire quickly spreads over the entire surface of the oil. The burning gasoline vapor raised the temperature of the fuel oil to its flash point.

The rate of burning is governed by the surface area. Only the fuel coming into contact with the oxygen in the air is consumed. The greater the surface area, the more readily the oxygen reaches the vapor. The surface area of a material in proportion to its volume directly affects its flammability.For instance, it is difficult to light a log with a match. The match merely chars a spot on the log, which absorbs the heat of the small flame. But a match will easily light a handful of shavings, and the burning shavings will easily kindle a log.

An ashtray is dumped into a wastebasket at bedtime. A live cigarette lands on some paper scraps. The cigarette is hot enough to both vaporize the paper and ignite the paper vapor. Paper has a very low ignition temperature. With oxygen present, this vapor will flash even from the heat of a cigarette.

Because of the large surface area of scraps in a wastebasket, this fire will really get going quickly.

It can easily kindle any nearby furniture, a rug, or even the floor. At this point the fire accelerates in intensity, building up higher temperatures, causing more and different vapors to be given off. The heat of the flames causes the hot air to rise, which in turn draws in more air, with more oxygen to combine with the increasing vapors and feed the growing flames.

It does not necessarily take a cigarette to start a fire. Oily rags or even water-wet rags can oxidize and generate enough heat to explode or kindle themselves into an active fire. This effect is speeded up when the heat of a nearby furnace or heater or even a hot-water pipe adds heat to the material. The added temperature speeds up the vaporization. If they cannot escape, the vapor trapped in the rags accumulates and builds up a higher temperature. In time the ignition temperature of the rags is reached and a fire is born. In contact with other materials, like a nearby can of paint or thinner, these additional vapors mix with the oxygen in the air more quickly with the increased heat until they make contact with the flame and ignite. Because of the larger surface area of paint or oil or thinner, exploded from a container, the spread of the fire is now unrestricted.

An intense fire in one room can transmit enough heat through walls, floors, ceilings, and crevices to raise the temperature of materials in an adjoining room to their ignition temperature and into flame, even without the transmission of spark or flame. Paint, in particular, is highly flammable because of its oil and resin content. Heat transmitted from another room will cause the paint to blister and scorch. When the paint reaches its ignition temperature, it will flash and burn and ignite the entire room.

In order for it to take on its so-called dynamite status (one gallon equals sixty-six sticks), gasoline must either leak or vaporize; it must escape from its container and combine with the right amount of oxygen from the air. The storage of gasoline, within your house, your garage, or the trunk of your car, is literally a dynamite situation. Even in remote storage, it remains a threat to its surroundings because of its explosive power.

Although gasoline is called a flammable liquid, it is its ever-abundant vapor that burns and not the liquid itself. In a very real sense, this vapor can act to gasoline in very much the same way as a fuse acts to an explosive charge. But this gasoline-vapor fuse is a lot quicker and a lot wilder. In other words, gasoline is deadly and should not be stored in the house.

The less volatile but also dangerous vapors from paints, thinners, and kerosene will join gasoline vapor if they are able. Vapors from this common type of liquids are generally in the range of being about 2½ times heavier than air. Natural and bottled gas, on the other hand, are lighter than air and will rise. Between the two types of gas, your cellar or garage or kitchen could fill up completely with a deadly mixture of vapors in a full-cocked state of readiness.

If you have natural or bottled gas in your house, you also have at least one pilot light somewhere in the house, an open flame. Chances are one is in your water heater, another in your oven, and at least one more atop your stove. You may also have another pilot light in your gas clothes dryer, your gas air conditioner, even your furnace.

Even though you have no pilot light if your furnace is oil fired, your hot-water heater is electric, or you have an electric range, any electric appliance can supply a spark to do the flame's job. Anything electrical can short-circuit or spark. Even static electricity in cold, dry weather can zap out a hell of a spark. An instant-on TV set is always on, always hot.

That's all it takes.

What was the recipe for fire? Fuel, oxygen, and ignition temperature.

Even an oil-fired furnace or water heater has a flame going in it as well as the flash, or spark, or explosion that kicks it on. You can't be sure the firebox is that tightly closed against the outside air. Delayed ignition or an improper air-fuel mixture can cause oil-burner back draft—it sprays a couple of times and doesn't ignite. When it does ignite, it gets more vaporization than it needs and Boom!—there is an explosion that can shoot out into a room. If the room is in a state of readiness, there you go.

Ignited, or unignited, these vapors could readily go or be drawn into your air conditioner, hot-air duct, or any internal circulation system in your house and very quickly be spread throughout the house. They then could ignite on a distant spark or flame and blow it all sky high.

These flammable vapors travel out. When they ignite at a pilot light, the explosion starts there and then travels back to the original container at a rate of sixteen feet per second, and then most likely that container will blow. If it is gasoline, paint thinner, cleaning fluid, varnish, lacquer, or alcohol, it will blow.

These vapors don't even have to explode to kill you. You might not even be able to detect some of them through normal breathing, particularly if you are asleep. If you do detect any of these vapors, you should react the same as if you smelled smoke.

Electricity does not have to spark to ignite another material. Electricity can do its number all by itself without the presence of highly volatile vapors. A short circuit can ignite a wall, floor, or ceiling directly, without an intermediate kindling agent. Basically, the only difference between a red-hot heating coil and two parallel strands of electric wires with full current in them is the insulation around the latter. If a rat gnaws on the insulation, if it cracks open from age or use or flexing, or if it burns through because some idiot has bypassed his fuses with a penny or taped the circuit breakers on, there's enough heat in those wires touching to start anything in the house burning. It could merely be an extension cord worn thin under a carpet or a door. Or a nicked wire. Or an electrical octopus, as they call those overloaded extension cords. An ordinary electric iron can burn its way through its ironing board or a floor or a table if it's forgotten at bedtime.

Fireplaces are a blessing, but that little fire you leave behind can kill you in one or two ways: If you close the fireplace off, the fumes from incomplete combustion could kill you with no fire or heat. Or it could shoot a spark to a rug, drapery, or stuffed chair or sofa. Likewise coal fires are dangerous, especially the gases they produce. Charcoal must never be used to cook or heat with inside the house. It is an absolute carbon monoxide generator.

Anything burning in any way with inadequate ventilation can kill you while you sleep. Last-minute house cleaning before going off to bed can do it—the messy ashtray into the messy wastebasket.

A window fan or air conditioner can overheat and start a fire at night, while masking its own distress sounds.

Smoking in bed. Ah . . . there's a biggie.

I'm not trying to scare you to death. And I'm in no way sermonizing fire prevention. That will have to wait.

I am trying to tell you what fire is, how it can start, what it will do if given the chance. I am trying to relate its characteristics to your home, especially to the situation in that home just as you turn off the lights to go to sleep for eight hours.

CHAPTER 7

Plan Now

In preplanning your escape, don't take anything for granted. Check everything out, test everything and, above all, be sure that common sense prevails. Keep in mind that survival is your single most important objective.

There is the question of windows. If you are on a low enough floor, a window that will open is an obvious escape route. Jumping out a first-floor window is not a particularly dangerous thing to do, but even though you're on a higher floor, that window should still be openable. You may need the air, you may need to signal from it, or you may need to retreat from the fire to a ledge or to an adjacent roof or tree.

If you're in a basement, you'd better make damned sure its window is large enough for an adult to get through—any sized adult.

Unless you have two workable exits, you'd better move or change bedrooms. I wouldn't sleep in a place I couldn't get out of, not for one night, not even for a nap.

So now you have got to choose not to sleep in a fire trap, a booby trap. You must now tackle that dilemma of hearing your children at night while still maintaining the safety of closed bedroom doors. You can decorate your bedrooms, you can arrange the furniture around this priority. But the problem has to be solved. Keep the children near to your bedroom at night. The younger they are, the closer they should be. And keep those doors closed. Make your sleeping arrangements that way from the very beginning, from the day you move in.

It is imperative that you preplan how to open all critical windows. (Why not all windows? You never know.) Make sure all of you —children, too—can open them. Your responsibility will automatically extend to those too young or too feeble. If they are in your care, they are your responsibility.

If some windows have been painted shut, you had better fix them. Get them working now, or you'll have to break them. Unstick that paint, get the painter back in to finish the job, call the landlord, or do it yourself. Make sure even the children can open them; show them how the locks work.

Is it a good idea to teach children to unlock and open windows? Most people don't want their children playing around with the windows. If the window is essential to them in their primary or secondary escape plan, either they have to learn how to do it or you'll have to include opening it for them in your personal escape plan.

Sooner or later you'll teach them how anyway. You'll have to be the judge of whether they're ready for it now. Otherwise, they are your responsibility when, and if, the fire comes. It could slow them down then, waiting for you to come; it could slow you down, getting there, when they could already be out and gone.

We find so many young bodies just inside the window. This may be caused by their not knowing how to open the window and waiting for some adult, who never comes, to do it for them. Forbidding a child to open a window in normal times can short-circuit his judgment at this critical time. It's your decision. You know your children. If they can handle it, you'd better train them to do it.

At the critical time, if the window won't open, even if it always has before, check the lock, then try the window again. If it won't budge, break it out. That's going to be rough explaining to the children. It's the same thing really. If that is the escape exit, don't let anything stand in your way.

In addition to arriving and going straight to the roof and chopping a hole, firemen also seem to break out all the windows routinely. Do you know why? For ventilation. But do you know why the windows? Because they're easy to break compared to a brick wall or wooden framing. If the house generally holds up, glass is comparatively cheap to repair when it comes to getting that house put back together.

So when you tell your child it's OK to break that glass if he has to in case of fire, don't have the cost of replacing the glass in the back of your mind when you tell him. That youngster will see the dollar signs spinning in your eyes and will take his instructions to be part no-no and that will interfere with his acceptance.

When you break the window, remember glass is sharp and dangerous. You have to be careful and quick. Pick any object you can handle, a chair, lamp, broom, anything you can poke or throw to break out the glass. If it's a chair, stand to one side and strike the chair through the window. If it's a broom handle, stand back and push it through as many times as you have to, to get the glass out.

If you're throwing something, throw it and then use something else to clear the glass out of the frame so you can get through. Broken glass itself is treacherous and deadly. Don't leave any big pieces hanging; don't break it out with your hands or even your feet. It's a good way to slice arteries and nerves.

In one fire we investigated, a man was half hanging out the window when we arrived. He was dead. We don't know whether he died from smoke inhalation or loss of blood. When he tried to get out the window, he sliced himself wide open, cut an artery in his upper body.

I would like to see more emphasis put on the proper operation of windows, especially storm windows. The more complicated the combination, the easier it should work. Young children could break the glass accidentally just trying to get the thing open. They would be better off breaking it out in the first place, from a distance, than having it break on them trying to get it open.

The glass above or below will be most important to you, depending on what you're after. If you want to get out in a hurry, open the bottom pane and let the upper be.

The upper pane will do more venting if it's just air you're after. If you're on an upper floor, you may want to break both of them. Break the lower so you can breathe air in. Break the upper to let the smoke out or to raise the smoke level in the room.

Now you may at first want just to clear the air and later you may decide you have to get out. *Clear that sill.* Clear all the glass guillotines out of both frames. Take anything soft—bedding, clothing, towels—and throw it over the bottom sash so you can climb out over it without cutting yourself to ribbons.

Screens can pose problems. Try to raise the screen. Be careful; remember you've probably got a lot of broken glass hung up in there. If the screen won't open, poke a hole in it with a chair, a broom, anything, and keep enlarging the hole until you can fit through. Or try to knock the whole screen frame out. A screen could end up trapping you. Keep them operable.

Screens are bad enough, but because of burglaries, lots of people are putting heavy security bars on the windows. You will never get out of some of them. Because of this problem (and crime *is* on the increase), fire codes now require that these security grills have safety releases on the inside and that they be hinged instead of bolted in.

If you've got bars at the window, you had better find out what to do about them. If they are bolted tight, you won't use that window.

While we're on that subject, basement windows, metal-frame windows with small panes, awning or jalousie windows can block you in, too. And you won't break out safety glass, unbreakable glass, or plexiglas. (You'd better take all of this into account in your preplan. And solve it.) If you've got only one window, it had better not have an air conditioner or a window fan in it.

The situation or condition in your home could change too, as remodeling and things of that nature go on. Don't let a painter paint you in or a carpenter nail you in or scaffold you in.

You'd better get all this into your preplanning, flexibly, if necessary—and keep it up to date with the actual conditions however they may change.

A lot of people think that if they have to go through a lot of smoke, they should put a towel over their face. Probably a wet towel will help in some minor way. But don't let it slow you down; if you're breaking out a window or close to getting out and have to use your hands, you'll have to decide what's going to do you the most good.

A wet towel over your nose or mouth or even a handkerchief will keep some of the irritants in the smoke out of your throat and lungs. It might keep you cooler, too. It won't keep the gases out.

You're probably better off stuffing that towel under the door to hold back some of the smoke. A wet towel, or anything made of cloth, is good to stuff at the bottom or along any cracks of a door to keep smoke from pouring in. Do that; it will give you time. But if you happen to have two towels, or three, one of them could help you breathe if it won't hinder you too much.

As you move about in a fire emergency, the most important thing of all is to keep as calm as you can; don't overexert yourself. If there is smoke, exert yourself as little as possible. The faster you breathe, the more air you take in. The more air you take in, the more smoke and poisonous gases you take in and the more oxygen you need to keep going.

To make this point clear: Carbon monoxide is a product of incomplete combustion. Each particle of it is a molecule consisting of an oxygen atom and a carbon atom. In complete combustion or oxidation that carbon atom would be complete with two oxygen atoms clinging to it, but that would be CO_2, carbon dioxide. Carbon monoxide demands that other oxygen atom wherever it can get it. In your lungs it takes oxygen from you to complete its circuit. It becomes whole, satisfied, when it gets its second oxygen atom. But each oxygen atom it takes comes from you, your lungs, your bloodstream. So the more carbon monoxide you breathe in, the more oxygen you lose.

At the same time, when a fire is burning near you the fire is combining oxygen with carbon and other elements for its oxidation. Fire *is* rapid oxidation; that is part of its definition. Burning *is* oxidation.

All of which means there is less oxygen in the air around you than there is, for instance, in the normal air outside, away from the fire.

So, in the first place, the air you breathe in the area of your fire has less and less oxygen in it. The more it burns, the higher percentage of carbon monoxide and other "poisonous" gases there is in it. The more of these there are in the already oxygen-depleted air you breathe, the more dangerous it is for you to breathe it.

So get the hell out. Carbon monoxide alone could keep you from getting out if you hesitate or delay.

You need 16 per cent oxygen in the air you breathe to function normally. Anything short of that will warp your judgment and instincts for survival. Anything far short of that will have a direct oxygen-starvation effect on your brain and you will lose consciousness and then die.

This is one of the reasons firemen use oxygen directly to resuscitate asphyxiation victims. Sometimes it can revive them. Sometimes, they are technically revived but are already suffering from massive, irreversible brain damage.

So back to the windows: That's where the air is; that's where the oxygen is. Doors are the best means of escape, but windows are better than nothing. Take all the advantages of that window that you can; the arriving fire fighter will focus his attention on it. Therefore, consistent with your limit of exertion, throw things out the window, break things, bang on something, wave a towel . . . anything that will help. It's the wrong time to be shy or embarrassed.

Turning on the lights is a good signal if they are still working. It will help someone see you; it will help them see inside. A flashlight is a useful thing to have in the bedroom. If you can't get out, you can signal for help with it. It can help you find your way in the dark, but remember you should know your way around in the dark. It's your home and you should practice getting around in the dark as part of your preplan.

You say you sleep naked? Then get out naked! You have little time. Don't start looking for clothes at this stage of the game. You are a lot better off first trying to block smoke or heat that is coming in. That's another thing. If you must pass through, or close to, flames you are better off naked anyway. Clothed bodies of victims, alive and dead, show plainly the pattern of the clothes they were wearing, like a horrible tracing. Usually the clothed areas have the deepest burns, and often the surrounding burns are the result of the ignition of flammable clothing.

Burns are of varying degrees of seriousness. First-degree burns are nothing more than reddened skin, similar to sunburn. Second-degree burns are reddened skin with blisters. Deep-seated second-degree burns are the same but more serious; they require a temporary cover but no grafting. Third-degree burns mean total destruction of the skin, nerve endings, and pores; they cannot regenerate, and the skin must be replaced by grafting from elsewhere on the body or from a donor.

When smoke awakens you, proceed according to your preplan. Roll out of that bed. Don't stand up in the heat and smoke and gases. Whether you've got a mattress fire in your room or the whole house is on fire outside your door, hit the floor. Stay low and proceed according to plan. You can get and keep your bearings better on the floor, particularly when you are dizzy, disoriented, and groggy, and when so many possible hazards await you. If you are disoriented, work your way around the wall by crawling; use the wall as a guideline, a safety rope to find the door, the window, whatever your first preplanned target is. Even if you know your room well, this will reinforce your preconditioning, give you time to cope with your panic and disorientation by giving you something positive and constructive to focus on while you get control of yourself. And it won't be wasted time; you're going in the right direction.

Your mental conditioning has a lot to do with your safety at this point. Here's where the value of all that preplanning, prethinking, predigesting, and practicing bear fruit if you have really done your homework.

In the prefire family planning, every member should be responsible for everybody else's planning; make the planning work by getting the whole family fully involved and deeply concerned so that you arrive finally at the best plan possible.

In a real fire, each person is responsible for his own personal escape, for his own life raft, so to speak. But it is the family responsibility to see that every single life raft is the very best possible *before* the fire.

To carry the analogy a little further, two persons can sink a life raft made for one person. When two people sink a one-man raft, they can end up drowning each other in the fire much the same as two panicked people tend to drown each other by grabbing too hard and sinking each other in their unprepared desperation.

By placing responsibility in this preplanning, you will get the whole family more involved creatively in what they are trying to do, in their own salvation. It will make the experience more vivid and meaningful. And that includes their conception of the fire itself.

In any plan, you have to establish a boss. It is not necessary that the head of the household be the boss in the prefire plan. In our fire-prevention and survival programs with the public schools, we have found excellent receptivity in the sixth-grade student, the twelve-year-old. For some reason this seems to be the most effective age to communicate safety practices and fire prevention; at that age, the young person is enthusiastic to carry the message home, to take charge of the family fire safety, to persuade, to be the "boss," and usually with no conflict from his parents. Even the most paternal or maternal family can tolerate and accept this direction, because there is no conflict in overall authority.

There has got to be an alternative boss and a chain of command. Everybody has to know his part in the plan and comprehend the part of everyone else. It takes a boss to weave it all together into a full family plan.

One person has to grasp the gravity and level of realization needed to make a plan work. It is not a game and should not be made to seem a game. When your fire comes, it might still have the aspect of game about it and that could be deadly.

It is one job of the boss to help modify how the others react. The boss must be the model for the others, whether head of the household or sixth-grader, and must be the best one to lead the reaction and response of the others. If he makes a mistake, it may kill somebody.

The greatest thing to overcome is apathy, apathy in people to whom fire has never happened, who don't even want to think about it. It takes someone who is "turned on" to turn the others on, to get their attention. There is nothing dull or boring about fire. Quite to the contrary. Somebody in the family has got to get the whole family to think.

The ultimate objective of any evacuation, regardless of how it is executed, is to get to the outside of the building, to a predetermined location, so everybody can be accounted for.

This place, your assembly point, has to be far enough from the house so that you are not further endangered by the fire; yet you don't want your assembly point to be so far away that members of the family, perhaps dazed, can get lost and not show up at the point.

The assembly point should not be too difficult or too far to reach easily. It should be simple, logical, and appropriate so that none will forget it in his ordeal. When young children are involved, the dangers of crossing streets must be taken into account.

Should the assembly point be in the front or the back of the building? In most cases, it should be at the front of the house. The fire department will normally arrive at the front of the house, and it is the best vantage point to account all present to the firemen and to volunteer vital information. There are obviously conditions, however, that call for an assembly point to the rear. With families living in midblock in rowhouses or duplexes or large single homes, where young children sleep in a rear bedroom, it may be necessary for them to exit independently from the rear. If the parents escape from the front of the house, it would be abnormal for them to wait in the front for their children. It would be much simpler for the parents then to go to the rear to meet their children. The route this movement takes should be preplanned to eliminate the risk of parents and children circling the block the same way, missing each other.

When the fire department arrives at the scene, the first thing they try to find out is: *Is there anybody in the house?* Now, what you are most interested in, before even notifying the fire department or advising them on their arrival if someone else has notified them, is the total and immediate accountability of *all* your family. This family accountability is your most important task.

So, if parents and children exit at different sides of the building, the parents should proceed to the children, wherever they are, from front or back, and all assemble there according to preplan.

This means a back *and* a front assembly point.

If both parents go to the rear to meet the children, then when the family is all together, notify the fire department and await their arrival at a *definite* assembly point . . . the safe point.

It is the safe point because each knows then that all are safe and can do no more until help arrives.

Now where should these assembly points be? You pick them; you know more about that than I do; the lamppost, the mailbox, a tree, a neighbor's steps, whatever promises to work best for you.

In fact, a neighbor could be part of your broader preplanning by mutual agreement. Use of their phone would come in very handy by reporting the fire.

By involving your neighbor in your prefire plan, you may provoke him into considering his own, possibly a consideration he had never before entertained.

The neighbor's participation could be especially helpful to a bashful five-year-old, who has to decide whether to ring a neighbor's doorbell in the middle of the night.

If you move, whether from one town to another, or only from one street to another, you must immediately dig in and develop a new fire evacuation plan. Once again it is the responsibility of the head of the household (or whoever is fire boss) to establish what procedure should be taken in the new habitat in the event of fire, smoke, explosion, or other emergency when it is time to bail out.

If you had good fire planning in your home, you will find it simpler this time; so attack this major problem when you first move in, even before the furniture gets put in its final position. Learn the ins and outs immediately—which doors, which windows, which hazards. Check out the neighborhood, the fire department, the fire alarm boxes.

Install that detector right away. Don't hesitate. Moving is a time when people lay their cigarettes on cardboard boxes "for a second" and forget to pick them up.

Those escape routes and assembly points should be checked out first thing. Get that room arrangement right, at the beginning.

Don't move into a fire trap; don't sign a lease on a landlord's promise to make repairs to provide your safety and security. Don't sign a lease unless you feel the building is at least up to the fire code.

This may seem a difficult situation when everything in your new habitat seems at its height of being strange and new. But it's better to get all that straightened out before you find yourself stuck in a fire trap. Later it could be very expensive to apply your terms.

If you're buying a house, check everything out for fire-code conformance as a condition of contract or be absolutely certain you can afford to bring the building up to standards, your standards as well, first thing.

Clean it up and clean it out. Chuck that packing material. Keep your exits free all the time.

There should be, as before, at least two clear routes for escape from any location in the household. Know all the alternates, all the combinations, all the possibilities.

Moving is a little like camping. Don't get lost in your new home. Always know the best ways out.

CHAPTER 8

Getting Help

OK. You have assembled. You're all out and together. Now one of you has to notify the fire department.

Do it yourself. You cannot afford to assume someone else has done it for you. Occasionally, there is a fire where everybody assumes someone else has made the call; it ends up that *no one has* and the house burns all the way down. In a way, the larger the fire the more obvious it could be to assume everybody has reported it.

In an urban area, you should preplan using the nearest local fire alarm box. When an alarm box is pulled, a light lights up and a bell rings in central communications, and a dispatch of fire trucks is immediately *sent to the location of that box.* You take it from there. Each of you should know where the nearest alarm box is. Each of you should already know how to use it. If you don't, go to your fire department; they'll show you. Chances are they'll have a practice box to demonstrate. The rule is stay at that box until the fire fighters arrive. But obviously your family is more important. If the fire can be seen from the box, the trucks will go right to it if they see it. Go back to your family. Have someone wait at the box to give directions, if possible. You have to decide what will cause less delay for the firemen to get to work.

In a large city, the fire alarm box is generally quickest but the other ways are almost as quick. In some areas, there is the Automatic Reporting Telephone, the ART, which is like having your own fire alarm box, your personal hot line to the fire department. Some security companies, along with, or instead of, their normal security alarms, can furnish a direct alarm to central communications, triggered by the same smoke or heat detector that wakes you up. They are expensive and they are not available in some areas; but wherever applicable, they are perfect for an invalid, for the elderly, someone recuperating in traction, or for large families with lots of small children. With these devices, your light flashes and the bell rings at the fire communications board automatically and the men and equipment are dispatched immediately from the nearest available station with no time wasted. It is the quickest way, but it is expensive.

You can telephone the fire department from a neighbor's house. *Do not telephone from your house!* That means going back in and you *must never do that!*

Obviously, everyone should keep the numbers of the fire department, the police, and ambulance service right at the telephone. If it's not there and you can't remember it for any reason, then call the operator. At one time that was advocated by the phone companies. Now they are trying to get away from it. I don't care what the phone company prefers. You dial that "O" and tell the operator you want to report a fire.

In some cities there is a planned push to make 911 the universal emergency number to call. There is a lot of resistance to that from the *rescuers*—fire people, police, and ambulance service—because right now, where it is being used, they get calls from people reporting everything from a lost dog to a lockout. That was not the original intent and its misuse can cause dangerous delays.

When you personally call in to report a fire, here is what to say, here is what they need to know—in this order.

Do not give your name first—that's last.

Be quick! Be concise! Be clear!

1. Say: *I want to report a fire. This is an emergency.*
2. Give the address, the exact location: the street number, and the nearest cross street, if you can remember it. The address is the most important single piece of information the dispatcher needs to know, because while you are adding to his information he can already sound the alarm in your nearest available station. First he needs to know where to dispatch men and equipment immediately; with your address he can now get the apparatus going. As you add to his information, he can dispatch other equipment, such as an ambulance.
3. Tell the operator or the dispatcher that it is an *inside* fire so they won't think it might be a trash fire, or dumpster fire, or a grass fire on a vacant lot.

4. Tell them what human life is involved. Tell them you all got out or tell them if someone is trapped or not accounted for; tell them if a neighbor's house has caught. Fire fighters respond to all fire reports, even false alarms, with the same commitment and sense of urgency. But let's face it, if they know someone is trapped, they will bust their asses. They will move heaven and earth; they will let nothing stand between them and your fire if they know it is a matter of life and death. They know that balance is entirely up to them.
5. Answer any questions you are asked. Can you give us better directions as to your location? What is your name? Is it your house? Is anyone injured? Repeat or verify.
6. Do not hang up until the operator or dispatcher tells you to. Some people call in and say: *My house is on fire!* and hang up. If you hang up on the operator without giving complete information, she may be able to relay the *telephone* location to the fire department and they in turn will dispatch to the location of the phone you called on. But this is a long shot. Don't depend on it! If you hang up on the fire department, they may have taped the call and set about to look for some clue to your location, but don't count on that either. If you hang up or become disconnected before giving the location of your fire, you can assume that call is lost and no one will respond until someone reports the location.

OK. Here's a recap:

1. Sound a box alarm if you can.
2. If you telephone, call the fire department directly, because it is quicker, it eliminates the middleman, and it eliminates repetition.
3. Use the operator as an alternate, your backup, or . . . think of her as your panic button.
4. Remember the old slogan: When in doubt, dial "O." It still works.

CHAPTER 9

Bailing Out

As a rule, the first-floor bedroom is as dangerous a place to sleep as any other floor. Here there is a curious combination of safety and underestimated danger. In terms of minimum distance to safety and abundance of quick exits, a first-floor bedroom seems to have all the advantages. In actual practice, however, it doesn't always work out that way. Beware of false security, which is a potent fire danger.

Three quarters of the sixty-three million occupied housing units in the United States are single-family houses: forty million one-family dwellings. An additional 6 per cent, almost six million, are two-family dwellings. Of the twelve thousand yearly deaths from fire, over six thousand occur in residences. Thirty-six hundred of these happen in the one- or two-family house. Half of these victims are under five and over sixty-five: approximately eighteen hundred children and elderly people.

Assuming that a very large percentage of these occupancies involve bedrooms on the first-floor level, I am going to deal with prefire planning and fire evacuation in terms of the special conditions affecting bedrooms and bedroom evacuation on or nearest ground level. After that, I will relate these factors to bedrooms located on second and third floors.

The various types of first-floor dwelling arrangements include ranch houses, first-floor apartments, mobile homes, camping and recreational trailers, houseboats, motels, and vacation homes.

There are a number of factors common to all of these lowest-level-bedroom locations:

1. There are usually plenty of windows at this level.
2. Most exterior doors are on this level.

3. Escape to the outside involves the least distance.
4. Escape from a window at this level is not a particularly dangerous thing to do. (You might sprain or break an ankle.)

Possibly the greatest hazard of a first-floor bedroom is its proximity to heating plants, hot-water heaters, major appliances, flammable storage, utility rooms, workshops, garages, trash storage, *and* the kitchen.

In one-family dwellings, there are more people who own their houses and fewer who rent. And in cases of ownership, there is a natural (but dangerous in case of fire) concern for property. That can be a serious hazard, even from a ground-level bedroom.

Let's take a look at ground-level windows. It is at those windows that one finds the unbreakable glass, the most screens, the most complicated and heavily fortified locking systems, and the highest concentration of grates and bars. All these security arrangements have the potential of becoming a fire trap, of slowing down one's escape.

A basement bedroom is probably the only general situation in which you would have to climb up a ladder to get out. And then you'd better hope you can get the window open or be able to squeeze through it. Basement windows are usually the smallest in the building and the most heavily fortified, and the least operable because of moisture and rust this close to the ground. Then there might be window wells and further grates to contend with. The window wells could be full of leaves or water or trash or snow. The window itself could be frozen shut.

You know about windows now. Make sure yours all work.

At first-floor level you will also find the highest concentration of utility lines: gas pipes, electrical wires and circuits, water lines, all sorts of wires and pipes and ductwork. This is also the part of a residence that usually houses the electrical service and panel, the hot-water heater, the furnace, washer and dryer, dishwasher, range and oven. Most of the pilot lights, or continuous ignition sources, are probably on the same level with all the stored flammables. Remember what fire is: fuel, air, and ignition heat. Remember also that fire prevention is simply a matter of keeping those three elements from coming together all at the same time. Thus, housed on the same level as your bedroom are the complete ingredients of a disastrous fire. That fire could be your greatest obstacle to escape.

Exterior doors are most plentiful at the ground level, but they are the most secured, in every way. There may be locks and bolts and chains involved. There may even be a deadly deadbolt or double cylinder lock, which must be unlocked from the inside to get out. Remember and consider this.

You should have the greatest possible number of exits and combinations of exits available to you on this level. Make sure the ones you need are usable. Don't count on a window just because it looks okay. Try it out. Don't plan to use anything that can't be used. Check it all out.

It may be the least distance to terra firma outside, but you'd better do a run-through on the actual step count. Just because you can touch grass with your hand or count the blades from inside doesn't mean you can plant your feet on it that quickly. Be a pragmatic pessimist in your prefire planning. Let your feet be your devil's advocate. It's a job for your feet; see how they feel about it.

It's far better to hit the front door forty feet away than to waste minutes finding out you can't get through that window or deciding you can't drop into a window well without breaking your body. Get out that front door while the getting's good.

The hazard of home ownership is somewhat subtle, perhaps even delicate. Despite the recent mass flight from one- and two-family dwellings to apartments, the basic one-family house remains the largest residence sector and the hardest hit by fire.

Now, here's the delicate, subtle part: Regardless of any particulars, ownership suggests affluence or at least the relative semblance of it. In this type of dwelling, where there is more ownership than rental, there is a higher propriatory self-interest. It is here that more people become more attached to things, and, therefore, more reluctant to let go of possessions at a deeper level. It is hard to abandon a house to fire—a house in which you have put twenty years of guts, blood, sweat and tears, your earnings, your life. That vested consideration, that emotional equity, is going to be a hell of an influence, even an antilife influence, to the guy who owns his own home that perhaps he and his wife built with their own hands. Somebody in his family may have to even physically pry him out of that house and to restrain him outside until he regains his senses. But if that house cannot be replaced, then it can be lived without. But nobody I know can live without his body.

Preplan the prefire disposition of anything in that house you love that much so that it will be safe from any fire. But do it now, *before* that fire.

In any living situation the invalid, the crippled, the incapacitated (temporarily or permanently), or the very old should sleep on the first floor. Because someone may have to carry that person out, and despite the perils of the ground floor, it is the level that permits the highest possibility of physically transporting another person to safety under the complex stress of the fire. This is true in nursing homes as well as private dwellings.

In your preplanning, wherever *you* sleep, that person in your household, who cannot physically effect his own escape, should be known in advance by the local fire department, along with his or her sleeping location. That person should have his own smoke detector and his own signaling device. That bedridden person could easily save your life with his alert signal.

If you *are* that person, I am talking to you now. You must have your own prefire and evacuation plan and apply it as fully as you are able. Regardless of your state or your condition, it applies to you as thoroughly as to anyone else.

The responsibility for saving your own life is your own more fully than it can ever be for anyone else. You must insist that the fire department know of your sleeping location and be notified of any change.

You will be rescued if someone can rescue you. But you must do everything within your power to escape, to rescue yourself, at least to aid your rescuer, for in so doing you may save his life as well.

If you cannot escape alone, you must enhance your rescue through any means you have. If you are bedridden, but have a telephone and can use it, call to notify the fire department, a neighbor, the police, or the operator. Get someone moving to effect your rescue and that of anyone else in the building.

You should have something to signal with, not just a panic button, but a whistle, gong, or horn. An aerosol boat horn works especially well.

I don't know who you are: an eighty-year-old woman, a nineteen-year-old boy in leg traction, an emphysemic man. Whatever, you are responsible for yourself. Someone *will* help you and rescue you *if they can.* People do it or attempt to do it all the time and some of them get killed just trying. But if they can't, you may be dead because you didn't do something you should have done, because you thought someone else was going to do it for you. If you can read this book, you can do more than just lie back and wait to be rescued or to die.

J.S.

This responsibility is what keeps people from dying in fires needlessly.

So, be alert, if you can; notify, if you can; escape, if you can.

If you can't escape, do everything possible to make yourself most accessible to rescue. That's your fire plan.

If it is preplanned for you to take your child out with you, you are telling that child to wait where he or she is until you come get him.

What if you can't come?

You have condemned that child at the very least to a dilemma that he can't think through and reason his way clear of. At the worst, your child will die of that dilemma, waiting for you to rescue him when he could have got out by himself—if you had realized this and given him that freedom in your preplan.

I don't know your child or your answer but I ask you to consider it this way: Between you and your child, you have to approach this together and fully enough so that both of you know very well what each can do, and both of you are going to know what you have to do and follow through on it. That's why practice and attention to this are so important.

Children grow. A plan that fit them six months ago may by now have been outgrown. They can do things better now and and can do it by themselves. You must recognize this when it comes about and teach the child to know it and modify the plan so that neither of you will be waiting within the fire for the other.

Escaping from a fire as a family is a delicate juggling act. What you are juggling is life, all your lives. It is so much easier for a solitary individual living alone to escape a fire. He has the best chance of all. It is your responsibility in your prefire planning to see that each member of your family—all of you—has the same chance of survival that the loner has, with no ties.

You must solve your ties in the prefire planning. You must free each other to the greatest possible chance for survival. That is the only way a whole family can survive its fire whole. Each of you must release each other to his own best path to personal fire survival.

Think of escape in terms of distance and time. Right now, think what is the quickest time you can make it to the street, the yard, in minutes or seconds. Think about practicing that after you have removed all the obstacles. Think about practicing it until you know it so well that you could do it in your sleep.

Now that you know every step to escape, count the steps. Let's say it's one hundred. Then you are one hundred steps away from escape from your fire.

Now, let's say you get up and panic instead of escape. You could use up those precious one hundred steps on your bedroom floor in a desperate, self-consuming tightening spiral right there in your own smoke-filled room with the fire beating at the door. And not go anywhere.

OK, you say, *we're beyond that. I know better than to panic now. I know how to get out. I can do it. I will do it when the time comes. I know how.*

But what about your child? Can he run down those steps? Or crawl out on the porch roof? Or streak out the front or back door? Or let himself down from his window?

Sure, you say, *he does it all the time. And a lot quicker than I can, even with practice.*

So what's your fire plan with him?

I'm going to take him out with me.

How many steps is it going to take you to go get him? Or have him come to you? Suppose you subtract those steps from the hundred steps it would take you to escape. How far will either of you be with the number of steps left after you have subtracted those steps you just spent?

Halfway to the door? Halfway down the steps? If he can get out by himself and you can get out and each of you knows the other will do it, then both of you should be out—clear of the fire.

Children can run like hell. Even if they're in a fire and know why they're running and know where they're running. Especially when they know *exactly* what to do. And that really is what prefire planning is all about. Responsibility and knowing exactly what to do—each of you—and doing it, without hesitation, when the time comes.

Concerning the invalid's preplan: If you have a partnership with that person and are responsible for that person's rescue, then whenever possible, that person should have a proportionate responsibility to you.

Think how that person would feel if he just lay there or sat there waiting for you, without doing anything because those were *your* instructions. What if you were killed coming in to rescue him and he was rescued and survived?

Imagine the guilt. Imagine how little that person would thereafter really want, desire to go on living.

Consider what steps, physical or otherwise (such as calling the fire department), the invalid can take to aid in his own rescue and perhaps that of other members of the family.

It is a matter of steps because it is a matter of time. I cannot tell you how fast fire is, how lightning quick is its acceleration. Seconds count; don't waste them—your own or anyone's.

Senile persons and infants are particular burdens; you assume them and preplan that responsibility. There are no two ways about it. It will be your cross to bear in your fire and you must carry it. If you preplan, you will make it, both of you.

In large, sprawling houses, more than one detector will be needed in your prefire planning. In general there should be one within thirty feet of each bedroom and an extra one inside the bedroom if you smoke. In a large house, signaling the alert may be a problem because of the distances involved between bedrooms. I would suggest an internal alarm system or at the very least some sort of panic button, like a loud bell or whistle, which the first one alerted can push to signal the others.

Seven million Americans live in 2.5 million mobile homes, isolated or in special communities. The number of fire deaths in these structures is not reliably reported, but the incidence seems disproportionately high. Mechanical failure or malfunction and construction, design and installation deficiencies are high, compared to conventional dwellings. Most of these fires are known to start in the kitchen and many of these kill, because the sleeping areas are so close to them. The fire spread is also rapid because of the extremely high flammability of the materials and finishes.

Any fire in an enclosed structure is a potential hazard to life. Here you have the greatest number of ignition sources—flames, pilot lights, fuel storage, and highly flammable construction and finishes—in the smallest possible space. In many of them, the greatest fire hazard—the kitchen—is between the bedrooms and the exit.

The prefire plan rules apply equally here. In the planned evacuation you must have two exits. Ideally, you should have an exterior door from every bedroom, but if you don't, make sure you have an operable window from every bedroom that you can pass through. Make sure no exits are padlocked or otherwise blocked, as by patio furniture.

Most mobile-home fires do not start in the bedroom but in the kitchen, so your detector should be high up, within thirty feet of every bedroom. If you smoke, you need two.

There is a larger chance of asphyxiation in mobile homes because of the tightly enclosed space in such structures. It is easy for even a small fire to completely exhaust the oxygen and extinguish itself; if you are asleep, you won't be able to breathe. Everything is speeded up here: smoke, heat, and fire all will accelerate rapidly.

No matter how proud you are of your mobile Home Sweet Home, to a fire it is a metal box full of flammables with poor exits and high risk. You are the closest you can be to living in an oven, and you can only hope no one turns it on.

Fire and smoke accelerate most rapidly here because of the small space involved and the method of its containment. All the ingredients of fire are as close together here as they will ever be.

Recreational trailers offer the same fire hazards as mobile homes, with the added danger of explosion from the pickup truck or car attached to the sleeping unit. Don't cook inside a trailer with wood or charcoal. Keep that hibachi entirely clear of the living or sleeping unit. These things look innocuous but they can be deadly, because carbon monoxide is deadly and you can't see it. Don't even let those charcoal fumes waft gently into your camper or tent.

Watch all your fuel sources and uses. Make sure everything is off and out when you go to sleep.

Smoking in bed is as deadly here as anywhere, or even more deadly since, being on vacation, you may say the hell with it and have a few more beers than you would be having at home. That little old smoke detector would be right at home here.

The possible danger of fire coming at you from the outside is high here, especially if you have any kind of campfire going.

Be sure the campfire is out. Don't park close to it at all. Make sure it or anyone else's doesn't catch the grass on fire. If an outside fire catches your camper, you would probably have zero exits. Make sure nothing around the camper leaks, not the least of which is your gasoline tank.

Check it all out in your prefire planning, even the immediate countryside that is temporarily your house. The door to your camper or tent is your bedroom door and probably your exterior door, but if the countryside is blazing, you may still have a way to go to escape the fire.

It would be good to have a second way out, but it's not likely the windows will accommodate you and, besides, the windows are probably only inches away from the door.

Preplan, alert all others and escape—whether in tents, sleeping on the ground, or in nearby structures.

Notify the fire department or, if that is impossible, do whatever you can to keep the fire from spreading into a general area fire. If you can't stop the fire from spreading, then save yourself, anyway you can. If you can't put the fire out, then it's not your job. Your job is to stay assembled and escape anyway you can and notify the fire department at your earliest chance.

See, I snuck one in on you. I said assemble and escape. You are now awake and alert as opposed to asleep and there the rules start to change. Now you can fight the fire if you think you can, if you have to. Now you can do some things differently because you are alert and not trying to escape to the outside. Now you are outside the fire and not fogged from sleep and paralyzed with sudden panic. Now, you can act like a full-fledged rational human because you are alert and clear of the fire.

A motel is primarily a one-story structure (sometimes higher) having outside exits from all guest rooms. Many motels are beyond effective range of public fire protection and water supply; more time is required for volunteer fire department response.

Chances are there is no fire detector in your motel room. That battery-operated model you have at home is light and compact enough to travel with. This motel room is your residence, if only for a night, and the conditions prevail as equally as if you were home. You are also a guest, so orient yourselves fully and quickly to all hazards and escape possibilities. I'm going to tie in boardinghouse and tourist accommodations here with no generality about their average safety. You will have to estimate the hazards of each situation yourself.

Now, here is your greatest advantage in a motel room: You have your own door to the outside. However, don't let that keep you from making sure your windows and any other doors are unblocked and operable.

Motels are poorly constructed in that the units do not have firewall stops between each other. Even if they do, the units may have continuous porches, wooden decks, and protruding roofs or canopies. That front door could be an impossible escape route if the outside woodwork catches.

Happily, there are usually no large or multiple losses of life in motel situations. Most motel fires are caused by guests in their rooms, usually a combination of alcohol and smoking in bed or in an overstuffed chair; sometimes fire is caused by poor judgment in dumping ashtrays. The other fires are the motel owner's fault. You should really feel safe in his place even for one night. If not, maybe you should look a little farther.

Because fire can be especially contagious in a motel, you have the responsiblity to alert all others once you are clear of the fire and to notify the fire department or personally insist that they be notified. It is not up to the owner to decide whether to report it.

Any fire should be reported, whether "extinguished" or otherwise. The fire department should make the final judgment that it is out. The motel owner may be a do-it-yourself handyman, but that doesn't make him a do-it-yourself fireman. In many cases, his handiwork is what caused the fire in the first place.

Under no circumstances should you extinguish a fire yourself, especially a mattress or stuffed-chair fire, and then check out of the motel without reporting it. Report it no matter how embarrassed you are; not doing so could kill someone. Reading about that fire in some distant coffee shop the next day could ruin the rest of your life.

Regardless of whose motel it is, it's your fire, your prefire plan, and your evacuation. Don't forget to alert your sleeping fellow travelers —even in the middle of the day—and notify the fire department the same as you would at home.

CHAPTER 10

To Jump or Not to Jump

Although many buildings of two and three stories are tenant-owned single-family dwellings, in this chapter we will be dealing with relatively small apartment buildings in which two or more families rent their dwellings from a landlord who may or may not reside on the premises. I am assuming a structure in which two to ten families live. It is further assumed that I have dealt with those living on the lowest level in this last chapter and that anyone who sleeps more or less regularly on the second or third level (including those who live in private houses) will gain additional insight into his particular problems by reading this chapter for broader application to his particular situation.

On the level of the second-floor and third-floor bedroom, the exterior fire escape appears, having not been present at the first-floor sleeping level and more or less disappearing as the height increases beyond the fourth floor.

On this level, windows are not only usable but highly useful as a lifesaving second exit or as an emergency exit. Above this level, their value as an escape route reduces itself to ledges, sills, and balconies for victim ventilation and rescue exposure only. From the second or third floor, it is possible to jump or lower yourself carefully without killing yourself. At this level, windows are usually large enough for even large adults to escape through and are relatively free of security features and installations that at a lower level increase the danger of escape.

At this level, the combination of large windows and exterior fire escapes creates a security problem for the tenant. Keeping burglars out tends to keep people in, a fact that is particularly dangerous during fire. In those households where fire alarms and burglar alarms are combined, it is an absolute requisite that the fire alarm override the burglar alarm function and be audibly different from it. In other words, in the event of fire, the fire security function is of supreme importance and its effectiveness must be fail/safe protected against any interference or malfunction due to the burglar alarm function combined with it in a single system. The burglar alarm must never interfere with or diminish the effectiveness of the fire alarm, whether triggered manually as an internal alarm or alert, or electronically triggered by a smoke or heat detector. Obviously, this in turn must not impair the security purpose of the burglar alarm.

The same thing goes for the fire escape and the security bars or grill mounted at each window or door exit to that fire escape. A fire escape that looks good, but can't be used because of bars that can't be opened instantly from the inside without tools during a fire, is as useless in practice as a fixed plate-glass window on the first floor that is unopenable without breaking it. Plate glass is very hard to break, and dangerous. The inherent false security in these three situations can easily cost you your life. You cannot afford to depend on anything which is not really there or usable. You must practice your prefire planning to be absolutely certain *in advance* that you can use what you count on using.

In a multiple-family dwelling, your exposure to the fire of others is increased and so is the exposure of your fire evacuation plan to changes and obstructions that can be brought about without your knowledge or consent by other tenants or by the landlord or his maintenance people. That is why preplanning requires periodical, frequent physical checks of all elements of your prefire plan.

Generally, the more families housed within the same building, the less each can do proportionately to preserve its own safety, and the more each finds its fate at the mercy of others whose habits and activities it cannot control.

This is one of the strong reasons why I am directing the burden of your safety to you directly and within your family to that basic unit of self-preservation and survival, the individual you. This is why I am not preaching fire prevention and this is why the messages on fire prevention directed to what everybody should do have failed. You cannot afford to assume that anybody else will do anything. When your fire starts, it doesn't matter a hill of beans how that fire started or where or why.

You can establish a tenant association in your building and other extremely valuable things in advance, but when your fire happens, it boils down to the same old basic situation: You can do only what is possible and you can use only what is available.

And so, with the "multiple-occupant factor," you are essentially in the same situation when your fire strikes as you were before and always will be. Preplanning includes prevention but when the fire happens, prevention is irrelevant. Preplan your escape.

There is one single factor that makes the bedroom location on the second- or third-floor level distinct from any other situation, and that is the possibility of jumping or escaping through a window by means of ladder, rope, or other device. At this level, but not lower or higher, you may have to consider the possibility of jumping to escape via your secondary evacuation route.

Obviously, if you are on the first floor, and you are talking about a three-, four-, five- or six-foot drop to the ground or garbage can or bushes, that's a sprained or broken ankle at worst, unless you are very old or are stupid enough to do it headfirst. From this height, the full length of your body lowered from the sill by hand is not a jump, or even a drop, it is a touchdown! If you are very old, a six-inch drop or even a level-exit slip could break your hip and that is serious, but generally the probability of jumping does not occur at the first level.

At the fourth-floor level or higher, jumping is almost certain death.

So, here's what's involved in jumping.

Exhaust all alternatives first.

The first thing that comes to mind is the possibility of getting to a roof of any kind—above or below you—to get you farther from the fire. It's probably a long shot, but even a tree or a pipe on the side of your building might be better than nothing. You will have to judge that. You'd probably be better jumping than falling; you have more control over how and where you land.

Another thing that comes to mind very quickly is the safety ladder. There are a number of flaws to the safety ladder. It may not be properly attached to the floor or the sill or the wall. The screws or lags might come out as a result of rotten wood, flimsy screws, or a bed attachment. The grappling hook or connection might not hold and you could end up killing yourself because the thing collapses and you fall and break your head or you fall into the fire.

There is also a possibility that the ladder is made of materials that provide additional hazards. Hemp rope can be suffering from dry rot if it has been in storage for five or six years and has never been used or tested. It could suddenly fail you when you need it most.

Certain rope materials can burn like a wick or melt like a candle if they pass through a flame, like a blown-out lower window or a ground fire below you. On the other hand, chain ladders can get hot enough to burn you badly if they dangle long enough in a flame or close to intense heat.

In open space with no foothold on a wall, the very mechanics of balance on a safety ladder can be treacherous and tricky. If you do have a foothold, a wall to press against, to press your safety ladder against, you have to be careful about passing through window areas. If the room within that window is superheated, the window you have to pass could blow at any moment from the heat and the pressure. If you touch that window with your body, passing over it, the slightest pressure could blow the glass. It could blind you or scorch you and you would let go and fall.

Climbing these ladders, up or down, is a hell of a thing. A child or even an adult, unless he is an acrobat or an athlete, is in for a rough trip. They are very difficult to use.

It is probably harder to go down these things than to climb up them. If you have the muscle and the endurance for it, it's better not to use your feet on the ladder itself but to use them to balance, to "walk down the wall" and use the grip of your hands only on the rungs. This calls for practice beforehand. Remember you're probably doing it in the dark.

And part of the preplanning is testing that ladder or rope or chain regularly. As part of your prefire plan, test it, practice using it, practice setting it up.

It may be practical to tie sheets together; it may for you be the best or only thing you can do. If you plan for this, practice before you surrender your life. Your full weight is going to be considerable and will you be able to locate and tie the sheets in the dark? Tie them to something you know will hold. If you tie them to a chair, make sure it is strong enough to hold the weight. Wedge it firmly so it won't slide.

Two sheets will reach only about eight feet if you have their anchor close to the window opening. But even if it doesn't reach the ground, it will take you down much closer so that your fall or drop won't be quite as far. Anything you can do to get closer to the ground is to your advantage. Even one sheet, if you must drop, is in

J.S.

your favor. In fact, if you do no more than ease yourself out feet first and lower yourself so you're hanging from your sill by your fingertips, it will lessen your total fall by five or six feet—your extended height. Your feet are that much closer to the ground.

You should take the best choice possible in where you land and how you land. Some of that is really still up to you. I'd rather fall on concrete paving or even steps than a picket fence or any fence for that matter.

Go for the best chance you've got. It may be you won't have to make it to the ground. The length of whatever extends your escape down might put you on a lower sill or ledge in a safer area. You might escape inside at this point, below the fire, into a safer room, possibly with a clear exit. You might luck into an outside ledge and stay relatively safe until help comes. But don't keep on trusting your weight to the sheet or towels or rope once you get your balance. Remember: The room you left is burning and your rope will burn too.

Children generally are much more adaptable than adults and obviously much more agile. It is easier for them to learn how to use a rope ladder or even go down a sheet than it is for an adult. So you may find in planning escape routes for the family that the children may have more alternatives than the adults.

If you have nothing to slide down on, and you must get out, it may be possible to throw pillows, blankets, clothes, or anything soft onto the ground in a pile and try to land on it. It might make the difference between breaking your leg or your back. Or you just might luck out.

Now this is a tough one, but you might be able to throw out a mattress and land on it, but you must consider the risks and they are also large. The exertion, in all that smoke and heat, might keel you over. Physically you might not be able to handle it; you've probably never tried to throw a mattress out a window. Firemen do it frequently during overhaul and salvage, and most of them would warn against it in your situation. Still it is a possibility. Also, you must think about where you'd be if you got it halfway out and got it hung up so you couldn't get it out or get it back in. You've seriously exerted yourself and you've blocked off your ventilation, but when things get that desperate it is often attempted. Sometimes it works. It is better than nothing.

The important point about planning to exit from windows is:

Don't put anything into the plan that is absolutely impractical or impossible for you to do, no matter how someone did it in the movies.

One way or another, you'd better check it out beforehand. I don't mean killing yourself by hanging on a sheet. But if the sheet seems a possible escape for the children, don't assume too easily that you can do it the same way unless it really comes to that as your very last choice. Even if the child can't climb down the thing, you can lower him and it will work for him, but obviously the child can't lower you.

If there's no one down there to untie or catch the child, you may have to lower him as far as possible and then let go and there goes the sheet anyway.

Unless there's a long drop for the child, that is if he is close to the ground, you'd probably better let him drop. The exertion of holding on could kill you or could strangle him. A window could blow out as he passed it.

Be careful where and how you drop him. Not on a picket fence. Not in a fire. You have some control over this and so may the child. You may be able to swing him to one side or the other. Between the two of you, pick the safest drop.

Again keep in mind balconies, setbacks, porch roofs, multilevel houses, roofs of adjacent houses, anyplace where you can get onto something farther away from the fire.

Beware of the hazards in that. Constantly consider your position. You may be perfectly safe staying right where you are and waiting to be rescued from that point. Don't run out on the roof and run to the edge and jump. Think. You may not have to jump. You may not have to leave that roof until someone rescues you.

Be careful. Don't blow your salvation by running into 220 volts or garrote yourself on a telephone wire. Watch out for roof scuppers or loose guttering, holes in the roof, rotten wood, or any other booby traps. Slate roofs are slippery as ice; they are also brittle and can break when you step on them and send you hurtling into space. Explore and investigate all these possible escape routes beforehand.

The extreme last-resort avenues of escape are dangerous. Escape to the emotional sanctuary of a closet or under a bed is deadly. But considering all sorts of exotic exits without fully testing all the basic exits can be equally deadly. Check out *all* your first resorts first.

When firemen plan a rescue, how to get the victims out of a burning building, how to get themselves out, their rule is to effect rescue by the least spectacular means available. Normally, this is the safest, most direct, and most effective way.

CHAPTER 11

Don't Jump

If your bedroom is on the fourth floor or higher, you have one less exit available for your escape than you have at a lower level. You cannot consider escaping through a window by your own devices as a possibility unless you have a usable exterior fire escape. By law and logic, you should have access to one in good working condition, one that is not itself a hazard. In very tall buildings, however, exterior fire escapes are not generally present. In a few situations, you may have access to adjacent structures or to your roof. Then you can continue your escape to a safer rooftop.

Lacking either, you must accept the fact that it is too high to jump. The chances of your surviving a fall from fourth-floor height—approximately thirty-five feet—are too low to consider. Despite recorded exceptions, if you jump from this height, you will probably kill yourself.

You are placing yourself in an extremely hazardous position if you choose to occupy a building of more than three stories when it is not protected by a functioning sprinkler system. It should also have an internal early-warning system and a fire escape that actually touches and can serve every dwelling unit in the building, including yours. If the choice of habitation is up to you, you may be very foolish to place your family in such jeopardy; residents of upper floors have an extremely higher chance of being trapped than those who live on a lower floor in the same building.

In such a situation, the sprinkler system is entirely the responsibility of the landlord to install, maintain, and have periodically tested. Make it your business to be assured that it will actually function properly in a fire and is not just hanging up there looking good.

Within a multifamily dwelling of this height there should be an internal fire alarm system of the sort that alarms *all* living units the instant *one* tenant signals a fire anywhere in the building. This alarm should automatically report the fire to the fire department. The installation, maintenance, and testing of this alarm system are the responsibility of the landlord and is normally required by law. Be sure you know it works. It in no way diminishes your need for your personal smoke detector or evacuation plan. You'd better hope the guy who sounds the general building alarm has one. Think of the precious minutes he will add to everyone's escape time (as well as his own) through having his early warning. Makes you wish everyone in the building had one, doesn't it?

An architect friend of mine and I were walking and happened to notice the same building at the same time and stopped to look at it, in particular the old, tinder-dry, dry-rotting, sagging wooden fire escape leaning against it. "Looks like two drunks holding each other up, doesn't it?" he said. If you have no sprinklers, no building alarm system, and you've got this kind of a fire escape, *move out!* If on the other hand, your building is served by a fire escape that currently passes modern codes, include it in your fire plan, and practice using it until all of you can use it with utmost ease and familiarity. Remember you may be going down it in the dark.

Older buildings of four floors or higher normally have an open stairwell as the only means of entrance and exit. Newer structures normally have an elevator and one or more internal stairwells, enclosed or open.

I want to focus first on the open stairwell situation, with no elevator. Escaping a fire down this open stairwell is equivalent to Santa Claus coming down your chimney with a fire in the fireplace. This is the fire's most natural, most direct way out; it is the way fire and smoke will go if it can and there is very little you can do to stop it. This is its chimney. If you live on an upper floor, you are going to be catching a fire whose intensity may be more than it is at its source.

I said it was impossible to expect to jump from the fourth floor or higher and live. Now, I'm going to say it is impossible to escape four flights or more down an open stairwell that is on fire. I don't mean the smoke from somebody's wastebasket; I mean a fire, roaring, blazing, furnace fire. That's what I mean by trapped.

If you are alerted early enough, before the stairwell becomes a furnace, sure you can make it. But once the carpets and the paneling and the wallpaper, and the varnish catch fire, forget it, don't even try. It will kill you.

At that point you'd better hope someone has called the firemen because it's up to them now. Stay behind your door, ventilate the room with the best air, and keep yelling "Fire" out that open window.

In some buildings like this you cannot preplan a second way out because the building doesn't have one. Time really counts here because the only thing you've got to save you until the fire fighters arrive is an early warning and a prefire plan with *one exit*. One mistake in your fire plan and you have *no exit*. Without a smoke detector, you probably have *no exit*.

As the fire grows it will push to the top of the open stairwell and meeting obstacles there, will begin to build back down. It is then that the pressure starts to build, pushing smoke and gases into your apartment from around the door and from all the cracks in your woodwork.

OK, let's not assume the worst. What should you have preplanned for this occasion? Your detector goes off . . .

The plan is based upon your escaping by that route that has the least obstacles and dangers, granted time enough to do it. It includes an expansion of the alert to include the other occupants of your building; this is your responsibility to them and theirs to you. It does not imply rescue.

1. Roll out of bed and check out the situation and yourself. Orient yourself.
2. Alert all others: first your family, then the general alarm, or start yelling fire and don't stop yelling, even when you get out. Keep yelling until everybody's yelling and getting out.
3. Test all doors and close each one behind you. Get out.
4. Assemble and account for all of your family. You probably have only one entrance, so there should be no problem between front and rear.
5. Notify the fire department.

Remember, your responsibility to alert other tenants does not involve you in rescue at this point. And the degree of general alerting you do should reflect the situation you find. If the fire is starting to lick at the bannister, yell like hell on the way out. Bang on some doors as you yell if there is time. In the vestibule, push all the buzzers.

It is your responsibility to hit the central alarm as early as you can. There is a slight cost of time to you in this general responsibility

150
J.S.

to other tenants outside your family, but it is far outweighed by the advantage of your personal early warning and the assumption that they have the same responsibility to you.

Now, the implications of this sharing is an expansion on the basic prefire planning, which I will take up when we discuss high-rise buildings.

We are not talking about fire prevention as such, but we can talk about escape prevention. The flower pots and clotheslines and bicycles and extra chairs and roller skates and baby walkers and baby carriages and motorcyles and janitorial supplies and ladders on the stairways, corridors, hallways, landings, basements, entrances, steps, and vestibules are escape prevention. You have an absolute right to an absolutely unobstructed exit. This is the landlord's responsibility; he must police it and enforce it; if he does not or will not, he should be himself policed.

If you are a tenant in such a building, you not only have the responsibility but you have the obligation and the right to scream like hell if your exit via that stairwell is violated in any way, manner, shape, or form by the landlord or any of the other tenants.

Call the fire department or the police department immediately if that obstruction is not removed after your initial request. Such obstructions are against the law and violate building, occupancy, health, and fire codes. This is no time to be a shrinking violet, because it can be literally a matter of life and death at some future time, perhaps tomorrow, when everybody in the building is going to depend on that stairway being clear.

On a broader level, prefire planning should extend to your insistence that your community have definite laws to protect multiple-family dwellings so that if there is an open stairwell, there *must* be a secondary exit for escape from fire. It's just one step beyond the landlord. You have the right to insist that your essential safety be protected by law.

Where such law does not exist, you would be a damned fool to leave yourself in this position, living on the fourth or fifth floor of a building that doesn't have a fire escape. If you are indeed in that position you really ought to get out of there and move where you will have a reasonable chance of survival.

Should you find yourself trapped, what you need is time. Try to determine where the fire seems to be coming from and where it seems to be going. Go to the front and look out; go to the back and look out, or side to side, whatever. If the flames are shooting out the front go to the rear and vice versa. Is the stairwell toward the front or the rear of the building? Get as far away as you can from the worst of the fire and its most obvious direction. Go to the room with a window farthest away from the fire. Make this your assembly point. Forget being in the front to signal the fire fighters for rescue; they'll be coming from all directions and that's a fact. They'll have plenty of light, and they'll be looking for you.

Close all doors between you and the fire. Check as many of these doors as possible top, bottom, and the largest cracks. Close all windows in the intervening room. Stuff your door, that one nearest you, with wet towels.

Keep low. Don't overexert yourselves. Open all windows, top and bottom. Breathe outside air.

Calm yourself and each other.

Wave a towel, a rag, a sheet, and keep it waving. Yell "fire" if it is appropriate; by now if you are in this extremity it is probably quite obvious.

The high-rise building, where your bedroom is four or more stories above ground level, is generally more modern, of more recent construction than the general open-stairwell type we have just considered. It is characterized by having one or more elevators and one or more closed stairwells.

If it has one closed stairwell, your fire plan has one exit, one way out. If it has two closed stairwells, you have two exits, two ways out. Because you cannot use the elevators as a fire escape. It is important that you know why you cannot use the elevator; it is the single most misunderstood aspect of high-rise prefire planning.

Here are the reasons why you can't use the elevators: They are dangerous, they may end up trapping you, the elevator mechanism may fail, there may be power failure, it could drop or fall.

It is possible you could get on an elevator above or below the fire and be transported through or even to the floor or floors of the fire, the elevator could hang up there and stop, the door could open and that would be it. The elevator shaft can act as a chimney or flue much the same as an open stairwell. It is an air shaft and can transmit fire from one floor to another and even to the elevator car itself. If the car should stop for any reason between floors or the door wouldn't open at a floor, the smoke could get you and you'd have no escape.

1 2 3 4 5 6 7
J.S.

If the elevator is operable, it must be completely available for fire services. You might ask: *If I can't use it, how can they, if it's so dangerous?* I might answer: They fight fires and that's dangerous too. Climbing eighteen or twenty flights with equipment is going to give that fire a big head start. You'd better let us get right up there the quickest way we can.

That walk up is a killer. Fighting fire demands endurance; those stairs drain endurance. Our truck ladders reach only one hundred feet or nine stories; our snorkel can't reach much higher. We have to get up in that building as fast as possible.

It's risky but we'll make that decision; we've done it before because we've had to. If we know where the fire seems to be contained, we ride up to within one or two floors below it and walk up the stairwell.

In most places the elevators are keyed so that the fire fighters can override the system and call the elevator for their exclusive use. I remember a thirty-seven-story building, the fire was on the sixth floor and there were thirty-one stories of solid smoke. People were standing in the smoke in the hallways at every floor. The car stopped at every floor signaled. This is dangerous for everyone involved.

People are creatures of habit. They came up in the elevator and by God they're going to go down in the elevator if they die doing it, and some do arrive dead. Some people will wait forever rather than change their habit and use the stairs.

Under no circumstances are you to use the elevator in the event of fire, unless specifically directed by a fire department official on the scene, regardless of the height you are at in an elevatored structure.

There is no such thing as a fireproof building, any more than there's an unsinkable ship. That includes the British Airways Terminal, the World Trade Center Building, the Empire State Building, the *Titanic*. High rises are not built of wood, but they are full of flammables, including wood, wall coverings, carpeting, and ceiling and floor tiles. They have many flues, "natural" drafts, for fire to travel such as horizontal ones: corridors, above suspended or false ceilings; vertical ones: elvator shafts, service and utility shafts, duplex stairwells with the spiral staircase from one floor to another through a hole in the ceiling, trash chutes, stairwells.

It would be logical to seal the building from floor to floor, but in actual practice, although this is attempted, it is hardly ever effected. Even if the fire doesn't travel far, the smoke still can. And as we know, smoke kills. A fire on any floor can permeate the entire building above it within a very short time.

OK, your smoke detector goes off. . . .

Whether the fire is inside or outside your apartment, follow the same preplanning. It's the same rules all over again.

Roll out of bed. Check it out. Orient yourself. Alert others.

Close doors behind you as you escape into the hallway, especially the door leading into the corridor. If there is an interior general alarm, hit it. If you hear it already ringing, obey it. Alert your neighbors as much as necessary. Move swiftly and carefully to the nearest stairwell. If the hall is crowded, be orderly, don't panic, and practice defensive escaping. No pets, valuables, or television sets.

Your aim is to get all the way down and out. This stairwell is the only way out of the building unless there are two or more stairwells. Once you are in the stairwell, do not run; take your time; make your way down. You are relatively safe in the stairwell, as safe as you can be under the circumstances. It is the best place you can be until you get out of the building.

Some buildings are so high that local instructions are that for all floors below a certain floor, residents are to go down; for all floors above a certain floor, residents are to go up. An area sophisticated enough to engineer and construct a high rise can be assumed to have sophistication enough to have an elaborate and workable helicopter-rescue plan. It may be that the planners feel downward and upward escape routes would be the safest procedure. That certainly would depend on what floor the fire was on. You certainly would not go up or down *through* a fire.

If you have not been otherwise instructed—and do not have to go through fire to do so—proceed *down* the stairwell for two reasons.

1. Once you are out, you are out.
2. You have gravity working for you. That climb up is a killer anyway. You will be in a higher state of anxiety than normal; it will be less taxing on you physically to walk down than to climb up.

You would not under normal circumstances proceed to the roof for rescue unless your downward passage has been blocked.

So, proceed on down and out onto the ground level (or upward out to the roof level) and assemble.

If the fire department is not there by then, you'd better make a point of calling them because something is really wrong. That something wrong may be that no one has reported the fire.

You should inspect your stairwell in your preplanning. Codes vary from place to place. A relatively safe stairwell is a sealed and separate chamber in a building for fire escape. Like any other fire escape, it should be absolutely free and clear of any "temporary" obstacles. If it is not, you know what you have to do.

It should be constructed of fire-rated materials, which means masonry and steel.

All doors should be positive latching, which means that if a fire builds up behind it on a floor and develops pressure, that pressure can't force the door open.

Each door should have an automatic door closer; after someone comes through, the door will close itself and latch positively.

Some building codes require a positive air pressure in the stairwell. In other words, you have a greater air pressure in the stairwell than you do on the floors so that the smoke cannot get into the stairwell.

Each floor should be numbered in the stairwell. You should know where you are at all times; firemen must know. From our point of view it is easy to lose count of floors going up or down; that can cause confusion and cost time.

No door—interior or exterior—in this passage between you and outside safety should be blocked at any time. No "temporary" locks, bolts, padlocks, or anything. There has to be some kind of slip bolt or panic bar you can use to get out at any time. Some buildings are equipped with locking devices that are tied into the fire alarm system and actuated when the fire alarm is triggered; in the event of a fire alarm these doors automatically unlock. I suppose this is all right, but they must be regularly inspected by fire people. I much prefer the manual panic bar, in some form. Imagine reaching the bottom of your stairwell at last, just a foot or two away from freedom, only to come to a dead end at a padlock someone has installed to protect you from vandals.

All interior stairwells should have an adequate fail/safe AC/DC emergency lighting system so there will be light in the event of a power failure. I've been in such a situation when the lights failed; believe me, it is not pleasant and it is not safe. It's like the Black Hole of Calcutta, with darkness, lots of people, smoke, heat, panic, uncertainty, hysteria. A single spark added to that powder keg could result in many people being trampled to death.

You should also preplan your response if someone should block open a firewell door during the course of a fire. That should not be done because it can render the whole stairwell impassable; it could be catastrophic if there is no alternate stairwell. Close that door the instant you see it open; unblock it and close it. The only justification for a stairwell door being open is if the fire fighters have absolutely no alternative to blocking it open themselves. We normally have our own standpipe connections within the building; our pumpers put the water into it under pressure, from the outside. Inside at the fire we hook our hoses into that pressurized water supply at the standpipe connections. For some reason I cannot comprehend, these interior standpipe connections are built into some stairwells (even in new buildings), which means fire fighters *must* block stairwell doors open with their hoses to fight the fire, thereby unavoidably letting smoke pour into the stairwell.

Basically, high rises are like vertical towns. Unlike real towns with fields and yards to run through if the town catches on fire and the roads are blocked, these tower towns have equivalent populations with only one or two roads—stairwells—to escape the fire.

The lower buildings are generally older, and there is no way you can make a new building out of an old one; you can insist by improving and enforcing codes so that they become safer, but you cannot insist that they become safe any more than you can make them new. Many of them are beyond the scope and teeth of any code or law because they would be condemned and they cannot be condemned until they are replaced. People must live somewhere, even if many of them die from their choice or the necessity to live in these places.

It's the building we are now constructing that *must have* these things I am going to list. Many of the older buildings are fifty years or so old; we hope the new ones will last as long. We will never be able to add safety to these new structures if it is not in them to start with. Because so many people live and sleep in just one of these vertical towns and because these high rises are our newest architecture, we must build them right from the very beginning. Many generations will live within them. We cannot allow apathy, indifference, greed, or "influence" to affect the built-in safety of any one of these buildings. We can't afford our death rate. We know how to build fire-safe high rises; if we can't afford to build them fire-safe, we can't afford to build them. This is one place where fire codes should have no double standard or discretionary enforcement built into them.

The original design should include:

Fire-stopped, fire-safe floors; top of the line fire-rated doors at all dwelling entrances and at every stairwell entrance; smoke- and heat-detector systems in every dwelling and every corridor linked directly to general building and fire department alarms.

An internal person-actuated alarm system linked to general building and fire department alarms.

An automatic sprinkler system in each dwelling unit and in all corridors linked by water-clock valve or gate to automatically trigger the general building alarm and notify the fire department when the water starts to flow.

Direct trigger linkage between any internal alarm system and all internal alarm systems *and* the fire department.

A top of the line standard of materials commensurate with the best-tested materials we can produce, from mortar joints to bedding, from draperies to pain. Maximal fire-rating rather than minimal; maximal application rather than minimal.

All of the requirements indicated for the stairwell.

Two fire-safe stairwells, *minimum.*

A total building fire plan co-ordinated and maintained by jointly a tenant association and the fire department. Such plans are beginning to occur more and more frequently. In some places it is a legal requirement. Increasingly it is coming into being, even where it is not required by law, because the tenants want it, they request it, they insist on it. It basically involves preplanning by the fire department and subsequent training of the tenants. It involves fire prevention, regular fire drills, prefire planning, fire evacuation, and fire suppression in a close working relationship between these enlightened tenants and their fire department. It also involves enforcement, because there are some basic laws in this that involve the right of one person not to have to suffer from the willful ignorance of another.

It involves a fire warden for each floor or level and a building fire warden who works directly with the fire chief and the most appropriate members of his department.

It involves appropriate training in inspection, prevention, evacuation, and the use of such basic fire-suppression equipment as the fire hose in the hallways and extinguishers.

Each such plan is designed specifically for each building and, in many city-owned properties, this program exists among interested tenants, security people, and the fire department.

At this point I don't think the landlord will do more than is required by law, and the insurance companies are not far behind the landlords. All tenants should want to create such programs and some are doing it. If enough of them do it, it will become mandatory, to everyone's advantage.

It seems impossible to convince the landlord—and this holds true in virtually every landlord situation—that it is to his best interest to protect his property, to have a prefire plan. He is not going to be there all the time, but surely some of the tenants are. If the tenants are not prewarned and they don't get out, his property is going to be destroyed. There should be some kind of an incentive or way we can make it clear to the landlord that it is to his advantage, even if we use the insurance company as a vehicle, to save his property. It would be in his interest to be sure the people are out there to save themselves and thus protect his property.

If anything, the landlord is more apathetic, generally, than the homeowner and somewhat less apathetic than the insurance companies. They at least have reacted in a minimal way by considering the possibility of reduced rates as a reward for certain fire-safety practices, including installation of a smoke detector by the homeowner.

Most insurance on an apartment building is strictly for the building itself. Most buildings are owned by groups of people and, in a great majority of the cases, by corporations outside the city. Undoubtedly, many insurance companies, which make their living from fire and life insurance, participate either in the mortgaging of apartment buildings or in the actual ownership of them. Why aren't they co-operating in programs to have tenants protect them?

CHAPTER 12

Who's in the Red Car?

It was almost three-thirty in the morning and you—the battalion chief—could feel it in the air; something was going to happen. There was a low constant garble coming from the watch desk as communications carried on its normal routine of moving the men and equipment necessary to keep the city's fire department alert.

You would think that after eleven years in the same bed you would be able to sink into some sort of semislumber but it just does not come that easily. It says clearly in the *Manual of Policy and Procedure* that the battalion chief has the authority and responsibility to do whatever is necessary to protect lives and property under any emergency within his command. I guess that's what is really at the root of the constant gnawing unrest.

In the adjacent dormitory lie eighteen fire fighters ready to spring to action at the sound of the first alarm. Each one has been specifically trained to act as part of a team in a situation where seconds lost could mean death to many. Each man realizes that, at any moment, he may be called upon to risk his life in a mad dash through the streets to meet an unknown emergency.

Maybe it is the "what if" that really gets to you. The men under your command are well trained. You feel confident that under the routine responses, their actions will come normally and smoothly. Yet the unpredictability of people and combinations of events never seem to let you feel the next alarm won't be an unforgettable experience. The whole thing is really about people, their strengths, weaknesses, and vulnerabilities. When a disaster strikes, it is men, women, and children reacting in their own ways as others come to their aid. You would never believe the ways in which most of us will react when our life or property is at stake.

The panic, fear, and terror felt by the victims are equaled in intensity by the pumping of the fire fighter's adrenalin as he rushes to do his job. It is an agonizing and difficult responsibility: It's impossible to know not only what people are going to do but what they are thinking and know it is going to affect their next move.

Nobody wrote it in the book of rules, but the battalion chief has to be able to motivate and lead his forces in such a way that the men will rise to 100 per cent of their efficiency each time the alarm sounds. In baseball if you get a hit one in three times, you are a superstar; in the fire department somebody's life depends upon your batting a thousand all of the time.

The street noises drift up through the second-story window as the street light casts shadows across the ceiling, just enough light to step into the boots and grab the turnout gear in one smooth motion. After years of doing the same thing, you really never remember the first twenty seconds it takes to get rolling, the jump out of bed, the dash across the room, and the leap into a hole in the floor as you grab the pole and slide to the apparatus floor. The apparatus roars to life and the next thing you know, the radio is blasting out the location and nature of the incident. You have learned to brace yourself as your red car accelerates into traffic followed by tons of apparatus and manpower, sirens sounding, air horns blasting, red lights flashing, and gears grinding. You are hurtling through the city too fast to stop; yet there is always the possibility that you are not going to get there soon enough. That is one of the initial risks; not only can you kill yourself and your men getting to the alarm, but every driver and pedestrian you pass en route stands a chance of becoming a statistic if they fail to yield to your right of way.

There are two great fears that never escape the emergency vehicle driver's thoughts as he moves through the streets: the dread of cross streets and the chance a driver may not hear or heed sirens and lights as men and metal race to their destination. During most alarms, emergency equipment will be coming from more than one location to the emergency. The last guy you want to meet at an intersection is another fire engine coming from your left or right, an engine whose siren blended into yours and gave no warning of an impending broadside collision. That's a risk you take every day; you court possible disaster, counting on the abilities and experience of other emergency vehicle drivers for safe delivery of the men and their equipment to the scene.

Confident as you are in your men, there is still that great uneasiness over what you have to protect; fire and emergency are a way of life to the fire fighter, but the people whose lives you have dedicated your life to saving just don't understand how fast, and tragically, fire can affect their lives. Fire seems to be something that always happens to the other guy; if only people would begin to realize that in the United States, on the average of more than once every hour, somebody (who thought it was going to happen to somebody else) is killed by fire. Most likely, he could have done something to save himself if he had taken the time to prepare for such a probability before tragedy struck.

It seems so much like a merry-go-round; if only you could stop fighting fires long enough to get the fire prevention rolling. Yet, just yesterday morning, the call came about a child playing with matches. After over an hour spent trying to reason with a four-year-old, you don't think you won. The child was scared of the uniform, but both parents work all day and the baby sitter says their kids are just impossible to watch all of the time. That call had come from a nurse at the hospital's clinic, and when I stopped by to thank her for her concern, I saw that the hospital's remodeling and renovation program was in full swing. Construction workers were mingling with hospital personnel. Again, another foul up of the local bureaucracy: The building permits were properly issued, but nobody told the fire department that roads were temporarily closed and halls, escape routes, and corridors were to be temporarily altered. I notified headquarters and called the first-alarm companies to the hospital for an on-site familiarization of current conditions. You can imagine the confusion if we had had a hospital fire with those changed conditions because some person had failed to follow the normal department notifications.

The hospital, by its nature, is very much oriented toward the well-being and safety of its patients. The entire staff knows exactly what to do in case of fire, yet hospitals and nursing homes always send chills up our spines because the potential for disaster is so great there.

The inspection report on the new twenty-two-story office building seemed to satisfy all of the existing fire-code requirements and was accompanied by the inspectors' recommendation to allow occupancy. Within the next week, I am going to have to approve this report knowing full well that our longest fire-truck ladders can reach only to the eighth floor of the building. But with new buildings, there are new techniques and systems to protect the building and its occupants. The fire department reviewed the design and system and now it was clearly the chief's responsibility to be sure his men know the workings and intricacies of the building. The new office building posed unique problems, even though it was far better than the old eight-story building with open stairwells that it replaced.

For the sixth time this month, we have received a complaint of a fire fighter invading the privacy of a citizen while making a routine fire-prevention inspection. The inspector returned to quarters frustrated and angered by the occupant's refusal to let him conduct a routine house inspection. As I suspected, it was another case of fear of authority. It turned out that the occupant was an elderly lady living alone, with only a pet dog as the companion, and had failed to have her dog license renewed. She was afraid the fireman would report her. This may seem strange, but many people feel that firemen are policemen and confuse their responsibilities and duties. I just wonder, in case of fire, if this woman would stop to rescue her dog before escaping her burning house. If she did, it might cost her her life.

It isn't often enough these days that you get a chance to spend a couple of hours on the street with the kids; there is so much you can learn from them. The days of mandatory fire drills with no explanation are gone; today's kids want and deserve explanations. You can see the sparkle in their eyes when you take the time to answer their questions; so many lasting impressions can be made if we would only take the time. Why is it the children you want so badly to get to on the street often don't come into your arms until you are trying to blow life back into them on the fireground?

I bring you some of the thoughts and concerns of the battalion chief not to create a hero but to give you some insight as to who the man in the red car is and his degree of dedication. But the fire department cannot help you until they know of the fire and this too is your responsibility. So each time you see or hear apparatus responding, somebody somewhere is already in trouble. Their chances of surviving may well depend upon how well they have prepared themselves for the initial stages of the emergency, especially until the arrival of the first fire rescue units.

CHAPTER 13

The Fire Panic

3:28 A.M. The entire household was sound asleep. They had talked about their fire, and they really did mean to plan for it except they didn't . . . and now it was too late. They didn't really do anything wrong or illegal. They didn't set the fire nor did they do anything to cause its spread or destruction; they just weren't ready. . . .

The house was neat and orderly, full of life and possessions accumlated by hard work and savings, much like your house or mine.

Here is what can (and usually does) happen when a serious fire strikes and you are not ready. The cellar was large and infrequently visited and had in it a nearly new gas water heater and a washer and dryer. Water, gas, and electric lines ran in various patterns up and down the walls and the length of the open joist ceiling. The cellar generally was quite warm, warmer than the third floor, cooler than the second. The family had a minor power struggle going because the house had no storm windows and the thermostat was generally manipulated by those sleeping on the second floor. They got too much heat before the third-floor sleepers could get warm. When the second-floor people were warm and comfortable, the third-floor people shivered from the cold drafts. Then one of them would walk down the carpeted stairwell, turn the thermostat to 75° so the temperature might reach 68° on the third floor. Then instead of opening windows, the second-floor sleepers would move the adjustment to 70° or 72° for relief; a compromise between heat and bad conscience. Around midnight they all went to bed warm under plenty of blankets on this very cold night.

The bedroom of the husband and wife was in the front of the third floor. Their room catches most of the wind and is the draftiest. The door was ajar to permit some of the warmer air from the hall to enter and to enable them to hear their children who slept in bedrooms on the second and third floors.

The heavy metal fire escape is strong and well attached to the house. It drops like half-held arms and elbows to the ground. At the bottom and directly under the lowest landing before the ground is a pile of split firewood and kindling. The edges of the steps to the basement were splintery and slightly crushed. It had been a bitch getting the motorcycle in for the winter.

3:29 A.M. For some unknown reason, the water heater malfunctioned and blew out the pilot light. The father, a carpenter, had installed the water heater himself; he was pleased; he had saved the $35 installation charge.

The wind outside hissed far more loudly than the unlit gas escaping into the room. The cellar, of all the rooms in the house, was least subject to drafts.

Meanwhile, the furnace roared on in overtime. Its metal front section glowed in the dark and through some small chinks around its edges it emitted little flashes of light like a tiny planetarium.

The explosion of the escaping gas at the furnace blew out the four panes of the back door and the panes in one of the side windows. It coated the cellar walls with an instant tongue of flames, but the house didn't catch fire until the motorcycle on its kickstand across the room from the furnace exploded. It had a half tank of gasoline, the cap was loose, and it blew the flaming gasoline over everything in sight. Although there was a heat buildup and the fire accelerated, there wasn't much smoke because the missing windows supplied plenty of air.

The heavy carpet caught immediately and began to eat its way up the open staircase. At this time the heavy logs in the firewood pile were roaring, the kindling already ashes.

The flames followed all the natural flues and ducts within the partitions around pipes and wiring as inevitably as water would have flowed down through them. The full height of the stairwell was filling with smoke. The banister, steps, walls, landings, and ceilings were beginning to smolder.

The doors at the second level fitted loosely and permitted heavy smoke to creep into the atmosphere at that level. The two girls who lived there slept on. The third-floor children's bedroom doors were tight and very little smoke squeezed in. A few ribbons around the top and bottom, and through the keyhole.

In the large front room, smoke began to curl up from the cracks around the baseboard like Medusa's snakes.

3:32 A.M., Four minutes had passed. The street was quiet, except for the bakery truck rounding the corner. The driver, an elderly man, thought he saw a flash and then a small cloud of smoke. He stopped in front of the row of parked cars and stared in horror. Fire terrified him; he was frozen. Coming to his senses, he began to honk and drove off to the all-night gas station two blocks away to report the fire. He went right by two illuminated fire alarm boxes. The truck driver roared into the station yelling something about a fire "up the street" to the half-awake attendent who went to the street and saw at a distance some smoke and an orange glow. He began to run to the fire right past the fire alarm box.

The husband awoke from the honking, instantly in fright, already wrestling almost bodily with the flood of fear and confusion within this sudden awakening. The wife snapped upright like a shotgun broken open. She began coughing and couldn't scream. He pulled her down and hugged her desperately for an instant. "Fire," he said to her. "Fire. This is it."

He slid off the bed onto the floor and pulled his wife after him, but she had already regained her wits. Although she could not yet speak, she signaled him with her fingers in his hand that she was all right.

He yelled "Fire" but started to cough. Smoke was increasing in the room. He tried to scream "Fire, Fire" but his voice was too flat and husky. The woman tried to scream . . . a gurgle came . . . then her voice broke loose and got a good one. She screamed "Fire, Fire, Fire" and at the same time they heard a banging from the back of the house, and before she could yell out, "Fire" again, recognized the sound of drums. The boy remembered his signal . . . he's awake . . . he's hitting his drum.

"Fire . . . Fire . . . Fire!"

"My mother!"

The boy had felt the door. Warm . . . not particularly hot. He tried it. Not bad. Just then he hears a "shoomph" and slammed his door quickly. *That* scared him the very first time. He started to panic. I've gotta get out. He opened his shades. Too far to jump.

Smoke began to pour under his door. He felt the door. OK so far. He opened it cautiously. A lot of smoke, hotter than the last time, but still better than jumping. The front of the hallway was hot and so smoky he couldn't see beyond the bathroom door.

He dropped to his knees and crawled past the bathroom and tried to look into the front room. Too smoky. He could feel the flames beyond the smoke.

Maybe OK, but the bathroom's closer.

He turned and snaked into the bathroom, then to the window in the dark. He rose and tried the window. Stuck! It wouldn't open. He thought about breaking the glass but tried the lock, unlocked it, and opened it quickly.

He knew the route. He had done it many times before when he wasn't supposed to. There wasn't much smoke out there. Most of it was crowding him from behind. But it was really dark. He reached his right hand out and grabbed the fire escape railing; he had a coughing fit but held on. Then he reached across the opening with his other hand and pulled himself out.

He knew it was three stories straight down to the woodpile but he knew every fingerhold over the railing and onto the grate. It was sharp and hot to his bare feet.

At the first landing his feet began to burn. Flames were coming up right through the grate from the woodpile. Now he could see the flames as well as feel them. He could see the flames shooting out the cellar window at the foot of the steps. At least he was out!

The husband and wife dragged themselves to the door under the worst of the smoke. It was already hard to see the outside light from the window and their first exit was toward the center of the house.

Near the stairway entrance door, the man urged his wife to stay on the floor. He took a breath, rose, and placed his hip against the door and very gently tried the latch. The door pushed back on him harder than he was pushing it.

He yelled, "Go back! I can't hold it."

Flames licked at his bare feet under the door and singed his arms. The door was open only a crack and he couldn't close it. He gave a push against it and used its force to dive backward as far as he could. The door exploded open and the hallway became a blast furnace instantly. On the floor he crawled back to the bedroom and his wife helped to drag him inside below the flames.

He screamed, "No! Not this way! Out the other way! The other way!" Together they pressed their bedroom door closed. The wife did not wait. She crawled to the other street-side window and tried the table lamp. There was no light. The wires had burned somewhere. She threw the lamp through the lower pane, then fumbled, found a chair and broke out the lower glass, then got on the floor to breathe.

The man crawled to her side and helped her to the window. "Can you climb down a rope?" he asked her.

"We don't have a rope," she said.

"Sheets and blankets . . . make one."

"Yes, yes," she said.

They stripped the bed and working in the dark with the smoke closing upon them, they tied the two sheets together.

By now smoke poured in under and over and around the door as the flames gnawed like rats upon the edges. The room was getting hotter.

He tied one end of the sheets to the nearest leg on the bed and checked the other knot.

"Breathe. Breathe," he said. And she moved to the window. She cut herself on the glass edges. He moved up beside her and raised the sash.

She used a blanket to clear the sill of broken glass.

"You all right?" he asked.

"Damned right I am." She made room for him on the sill and leaned out to gulp the clear cold air. They saw a figure running up the street yelling, "Fire! Fire! Fire!"

He played out the length of sheet. It was short of the ground. How far short straight down, he couldn't see.

She straddled the sill and then leaned in, held the lengthened sheet to test and then clenched it with her hands and slid slowly out of the man's sight.

"Can you reach the ground?"

"No, I'm at the end of it."

"How far?"

"Don't know; pretty far."

"Hang on," he said. "I'll lower you; there's time. I'm going to untie and lower you."

"No, I'll drop. You . . . . hurry."

He took a breath and crawled to the bed leg and began to work the knot loose, but the other end felt loose. She had let go. The gas station attendant had tried to catch her; they both were on the ground now a tangle of broken ankles and arms.

He was on a pile of blankets. He looked around the room. "It's bad." He dropped the blankets to the sidewalk below, hoping they wouldn't glide out too far from straight down.

"You clear?" he called. No answer.

He started down the rope of sheets. Smoke was shooting out the first-floor level at the end of the sheet. He could feel the heat on his back. When he reached the end of the sheet, the smoke blinded him.

He dropped and hit the blankets instead of glass and concrete and rolled out into the pavement sprawling.

He sat up and looked for her. He hurried to her wrapping the blanket around her. "The children!" she screamed.

"I've got to call the fire—"

"GO!"

"Call to the children; warn them!"

He couldn't remember where a firebox was but found a brick and threw it through the neighbor's front window, then ran up the steps. As he pressed on the buzzer, the neighbors yelled. "Coming, coming."

The neighbors opened the door.

"I've got to call the fire department!"

"I called them."

3:36 A.M. Then they heard the sirens. The battalion chief's car was followed by a legion of emergency apparatus all converging with no regard to one-way streets, coming from all directions, surrounding the house.

He ran to his wife.

She said, "The kids!"

"You tell them I'm OK. I'll run around and see. Tell them I'm OK."

He ran into his son at the corner. "Where's your sister?"

"Right behind me."

"Meet your mother, quick. Let her know."

The man met his youngest daughter coming from the alley, limping. "I fell down, Daddy." He lifted her with all the strength in the world.

The battalion chief was asking the wife, "Who is still inside?"

He called to the truck, "Second-floor front, quick. Skip the first and third." He ran to the radio. "Person or persons, second floor, use air masks, the building is fully involved." A ladder truck screamed past and rounded the corner.

"One here," the call came down. "Alive. She's moving. We've got her."

The fire officer shouted to the waiting ambulance. Two men scrambled down the ladder bearing a child. The two ambulance men took her and began to administer first aid to overcome the effects of asphyxiation. There was no sign of any external burns.

The ladder crew was already carrying ladders into the yard by the time the pumper was connected with the hydrant. They had to break the front gate down.

Flames were shooting almost straight out all rear windows, all floors, like so many giant torches.

"Person or persons, second floor."

"More ambulances!"

Before the hose line even reached the second floor, the search was well underway. Ladder men were crawling through the second floor, reaching, probing through the smoke and heat, under windows, beds, in closets. . . .

The water came on.

"Second floor!"

"Right."

"Cool it off."

"Ventilate."

More ladders appeared and men appeared upon them.

Water sprayed from one, then two, then three nozzles.

In fifteen minutes the fire was under control. Men were in the building as far as they could go on every level they could enter.

The fire fighters pressed deeper into the house.

On the second level, they found the other child, dead at an open bedroom doorway. The family cried, the friends cried, the fire fighters cried. The morning news told of the tragedy; it had happened again to somebody else.

CHAPTER 14

Should You Be a Hero?

Suppose all of you but one are out. That person is trapped or stuck at a window at a level a ladder can reach and that person has the choice between jumping and using the ladder or *something* you can provide. Then provide it and preplan for its provision if the fire fighters have not arrived.

There are two methods of rescue if the person trapped is conscious and able. Get your ladder or a neighbor's and throw it against the side of the building. Don't throw it against the window; you could hit the rescuee or shower yourself with broken glass. Then move it over to the window at a safe angle from the wall. If the victim can come down, hold the ladder firm and steady. If not, go up or get somebody to go up to assist or do the rescue if it is at all possible. Use your strength and your common sense.

It seems callous to say, but one victim is better than two victims; or to rephrase that statement: Two victims are worse than one victim.

If there is no ladder, throw something up that will help: a rope. If there is no rope, the other way to help is to put something down to break the fall: leaves, trash bags, clothing, bushes, a car seat, an empty basket, anything softer than the ground or concrete. Two of you on the ground could hold a raincoat or topcoat out tight as a landing net.

If it's a crucial situation and you can't find anything in time, you can try to break the victim's fall with your arms, if you're willing to break your arms. Don't try to catch him and don't let him land on you, your head, or your back. At the very least you can hold a light on the landing area and help him decide where, how, and when to jump.

Think about it this way: It could be the other way around. You could be the guy stuck up there. You would know what you would want or at the very least what to settle for to break your fall, what you'd prefer to that hard, abrupt stop.

After you have escaped, don't disappear. In actual practice, many survivors do disappear and come back on the scene later, often after the fire is under control or out. It is much better for us to hear directly from you than to get misinformation from your neighbors or needlessly overexpose ourselves to your fire finding out the hard way.

We have been assured by neighbors more than once that everybody was out only to find a body in there. We need more certainty than somebody saying, "Yeah, I think everybody is out."

In your fire I'd believe you more than anyone else, but I'd still check it out anyway.

Go directly to the first fireman to arrive and identify yourself. Don't wait for the fire chief. This man will need to know from you whether everybody is out and accounted for. He'll communicate that and act on that first. He'll take your word instantly if you tell him someone *is* still in there. If you tell him someone *isn't* in there, he will nonetheless respond as if someone *may be*.

Answer his questions quickly and fully. He might ask you what you know about origin, location, and description of the fire. He may ask about the interior room and floor plan. You should offer information about any special dangers you can think of, for instance, things that can explode or collapse. You know your house; he has never been in there. Remember: He is not the operator or dispatcher you talked to when you reported the fire. Tell it again. You are not going to help haul lines or hoses or get involved in rescue or anything like that. Tell him anything you think he needs to know about your house, your fire, that is not readily obvious to him. Then get back, stay out of the fire fighters' way and do not interfere in their operation.

In a rare instance, a fire fighter might ask you to lead him inside for human rescue if someone was still inside a broken-up or complex structure where he would have difficulty finding his way to the trapped victim. That would be a last resort, but you must make every attempt to fully inform him in such a case.

Again, don't disappear. Make the initial contact. Offer full information and get clear of it. This can save someone's life. That life could be a child's . . . or a fireman's. Both are precious.

CHAPTER 15

Save the People First

It should be clearly understood that the only responsibility that is really yours to take, all things considered, is for the life and safety of the occupants of the household.

Forget the pets: the hamsters, the gerbils, the dogs, and cats. My wife said to me: "*You always wanted to get rid of that cat. You never liked that cat.*" Irrelevant, my dear.

There is only one obligation: Get out and get help!

This is where the head of the household or fire "boss" may have to assert himself aggressively.

The kid screams, "*I want my dog! I want my dog!*" You're going to have to hold him back and apply your reason and your strength to his hysteria. The assembly point is no place to argue or even discuss the possibility of going back in—for anything. The answer is *no*, and it must be enforced.

We've had children and even adults go back in because they couldn't locate everyone at the assembly point. We've had them go back in for pets and valuables. This must not happen. That is what the assembly point is for; that's what the boss is for; that's what the preplan is for.

People get hung up on pets. Their pet is their family; that dog or cat is their only child. Give them a few weeks and they will get over that. There is a distinction between an animal and a child. The distinction is real and time proves that conclusively. Pets must not be part of the escape plan. If the pet chooses to follow you out or escapes on his own, so much the better; if he doesn't it's too bad, it really is, but that is the reality. The only way a pet can be considered is in the prefire planning. At that point, when you are at the room-arrangement level, you can and should build in some way for the animal to make its own escape. That must be the limit of your responsibility.

It is very difficult to explain to a child that he has lost a pet in a fire or his doggie has been hit by a car and killed, but in time that wound will heal for him as it has for all of us. But it is impossible to explain to a child that his brother or mother didn't make it back out of a house one of them re-entered to get his pet.

This is a dangerous situation, delicate, tricky, bordering on the fatal where action can be triggered by that sudden strong whim, that unpredictable impulse. It calls heavily on the command decision of the "boss" to manage and restrain the situation, to see that reason prevails.

At a recent multialarm apartment fire, we were racing up the stairwell—or trying to—getting our men and equipment up to the tenth floor to fight that fire. You wouldn't believe how many fur coats, jewelry boxes, parakeets, and cats were coming down. The stairwell was so cluttered that the occupants didn't know what to do and the firemen couldn't make their way through to fight their fire. It would have been orderly had it been just people getting out, but it was like climbing up through a clearance sale in a zoo. A few people had the sense to put their dogs on their balconies and get the hell out. My hat's off to those people.

We had to forcibly restrain an otherwise very dignified gentleman at another fire. He had a dog on the third floor and was damned determined to go back in to get it when we insisted we had to get the people out first. He would have none of that. We had a hell of a time holding him back, keeping him from going back into that building to get his dog. Finally, one of the fire fighters went up and got that dog, after the fire was out. He brought the dog down and I was happy to see the dog get out, but I was having some not-too-hidden thoughts about charging his master with obstruction, thoughts I did not pursue despite my inclination.

There are many cases where animals do escape and do survive. I remember one fire where the dog was in the basement and survived because the fire and flames rose upward, away from him. In another one, a fatal fire, in which the whole front of the house had collapsed, where there had been intense heat and smoke, a mother cat and her six kittens just breezed right out like nothing happened as soon as the fire was out.

I recall the first major fire I went to, a two alarm. The house was totally burned out. It was unoccupied at the time; the owners were out for the evening, but their dog was inside. He was killed by the fire: overcome by heat, gases and smoke, not burned, the same way people die. It was a fluffy, good-looking dog. We brought him out. There was more concern for that dead dog lying there on the grass in front of the house than there was for the fact that the people lost their *home* and all their possessions. Some of these people moaning did not even know at that point whether any people had been lost inside.

In case of fire, your pet has a chance but something less than half a chance. He would have about the same chance you have except he can't open doors. Hope and pray for him; but don't kill yourself trying to rescue him.

Pets must not be included in the evacuation plan—period.

The same is true for valuables.

Firemen are not going to loot your place. I state this bluntly and baldly because it needs to be said. If you don't believe it now, you never will, but you'd better accept it and you'd better act on it because it is your life.

Deal with your paranoia at the same time you soul search your responsibility in the preplanning. You must assume beyond question that the fire fighters responding to your fire are completely honorable.

Any goods, valuables, possessions in your home that survive the fire, heat, smoke, water, ventilation, overhaul, confinement, and extinguishment of your fire will be restored to you through the salvage responsibility we assume as part of our job.

Firemen are extraordinary men, dedicated to the profession of saving life, going in, putting out the fire, creating the least amount of damage under the circumstances. If you have a hell of a fire, you're going to have a hell of a mess. Because we're going to put in all the water we need to put that fire out. We're going to pull apart as many mattresses, ceilings, walls, and floors as we have to, to find and extinguish all of your fire.

We didn't start your fire but we are going to put it out.

And we don't have the inclination or the time to steal from you.

When we arrive, we are the law, we are the security . . . over the police, over the population . . . over you. There is no way we will permit your interference in our work extinguishing *your* fire.

There are no better housekeepers in the world than firemen. Look at their firehouses, look at their equipment, look at their persons. They're not making the mess, the wreckage. . . . Your fire did that.

And when they're done, they will help you clean it up. And that includes, at any point they can, the salvage and return to you of your valuables.

Your concern is before and after your fire. While they are in charge, it is not your concern.

CHAPTER 16

Fires Do Come Back

She awoke to find her entire apartment engulfed in flames. She was lucky; she survived it. She lost everything. Before we had it out, eight floors with 120 people had to be evacuated because of the smoke and the danger. But she wasn't even injured, except for minor smoke inhalation and a large dose of fear she'll never forget.

Never to be forgotten was the fire they thought was out. It was a sofa on the first floor; they put some water on it and went to sleep. Awoke to find the complete house on fire, all three stories. Of the three families, six people died. Four of them were children on the third floor.

Another man woke up because his mattress was burning. Instead of calling the fire department, he put it "out." He carried the mattress out to his back porch and went back in to sleep on his box spring. The mattress reignited, setting the wooden porch on fire. When we got there the man had collapsed on his window sill trying to get out. He was unconscious. He was lucky: The fire was spotted by a fire department ambulance crew returning from an emergency run. They had the equipment on board to revive him.

Another man removed a smoldering mattress to an adjoining room and left that window open to "kill" the smoke. This "extinguished" fire burned through the knotty-pine paneling and the door and came back to burn him to death.

Some people stuff the mattress into a window to "cool" it. Along comes the wind and fans it into flame and that's it.

There are hundreds of cases like these, across the country. If you or anyone in your home smokes, the chances that you are going to have a serious fire are considerably greater.

People who are lazy, tired, careless—or who are trying to hide the fact they had a fire—have killed an awful lot of innocent people when that second fire erupted.

At one apartment house, we got a call in the middle of the night from an occupant reporting an odor of burning rags.

We got the call about 2 A.M. and investigated. The chief sent the men along several floors to wake everybody. There was a thin haze even in the hallways. It wasn't much to see, but you could smell it.

We went through the building once. Everybody said, no, we don't have a fire; everything is fine.

I still smelled it and I knew there had to be a fire there. I knew the peculiar way that a mattress fire acts, that it can smolder for two hours and then burst into flames. I smelled *mattresss* and I was determined to find out where it was, so we went through again.

We went into one apartment. The guy acted suspiciously. We asked if we could come into each apartment and nobody said no, so we went into his apartment the second time because he acted a little funny. His bed was neatly made with the spread over it. There was no sign of the smoke being stronger here than in any of the other apartments, but there was a little water under the bed.

We took a closer look, and of course the guy was standing there saying he was wondering where the fire was, too; he was very concerned, even scared to go back to sleep. A fire fighter took the spread and threw it back, pulled the blankets down and then the top sheet. There was a big hole burned in that mattress, down through the bottom. We took the mattress out and tore it apart and found a smoldering fire inside it the size of a wasp's nest. It was still smoldering, because that's the way mattresses behave. And this guy couldn't wait for us to leave so he could go back to bed.

This is why anyone who smokes in the bedroom should have a second smoke detector. First of all, the smoke can kill you if you're counting on waking up in time. Second, for some reason, it is possible to burn to death on a mattress without waking up. Maybe the heat does it. Maybe you can take so much burning and then you wake up from the pain and go into shock and lose consciousness; then the smoke gets you, and that's it.

If you do wake up and your mattress or sofa or whatever is burning, roll out of bed. Stay low. Alert the others. Get out of the room. Get away from the fire as far as you have to—and then a little farther. Assemble. Call the fire department.

You may not have to go all the way out of the building since you know the fire to be contained, but stay clear of the smoke. Call the fire department. In a city, they can be there in a few minutes.

If you opt to put water on the mattress, know what the hell you are doing; watch out for that smoke. And let the fire department finish killing that fire. They know how to do it—and you don't.

They've seen hundreds of these fires. How many have you seen? And for God's sake don't go back to sleep until it's *all* over with.

CHAPTER 17

Children and Matches

How are you going to stop children from playing with matches? Can you stop children from playing with matches? Children and matches cause about seventy thousand fires each year in this country. These fires kill hundreds of people and injure thousands more. Many, many of these victims are the children themselves. Children playing with matches account for about 7 per cent of our yearly fire toll.

There are two types of children who play with matches. One is the normal, inquisitive child growing up and trying to learn more about himself and the world around him. Sometimes—especially between the ages of two and six—he may find out more about it than he really wants to know or is prepared to handle. With matches and fire, he is definitely biting off more than he can chew. This normal, inquisitive child can usually be reasoned with and explained to and shown how and why. Normally, if he has been satisfied that his questions have been answered, he will accept the information you have given him and that will be it.

But, over the age of six and at any age where the person persistently or periodically and secretly "plays" with matches, he is not simply "playing" or investigating the universe. He has a problem, often deep-seated, more often than not connected with other causes and symptoms of disease and rebellion or anger or intelligence or something I am not qualified to speak about in any detail.

What I can tell you about this type of problem is that it is your problem, too, be it your child, your neighbor's child, or a playmate of your child. His or her "playfulness" and your matches could kill you and your family. It could burn your house down. The child who persists in playing with fire is a potential killer. If this is your child, or one who comes in contact with your children or home, you should insist that this child be brought to the attention of someone trained to understand and help solve this problem, a psychologist, psychiatrist, social worker, someone who can take it several steps further than you can if your efforts have had little effect. You are going to have to take the initiative. Don't delay dealing with this problem; it is a serious threat to your life.

Obviously if there are no matches around, children can't play with them. If there is no gasoline around, it can't explode. I am not judging smoking cigarettes, cigars, or pipes. If you are a smoker, obviously you use matches or a lighter. I would strongly suggest that you control your matches or light as far as possible. Keep your matches out of the reach of children and you are in some control.

What do you do? How do you train a child to control himself in terms of the danger of matches? Do you make matches a *no-no*, a smack in the face or on the fingers the first time the child even shows an inclination to play with matches? If giving the kid a whack would definitely save a life right out from the beginning, to be sure you should give him a whack. But do you really know? I don't.

How do you handle it? Do you burn his fingers with it? It's like sex; you've got to tell them something about it or they'll find out for themselves. Unlike sex, this could kill them and you.

In educating the child many parents use some kind of a demonstration technique. I don't think it should include burning children's fingers to let them know what fire is. I think there are better demonstrations to make the point to a child. And I don't think you should limit the demonstration to all the bad things fire can do. They need to know most food can't be delicious if you don't cook it. It's a lot nicer to take a warm bath than a cold one. A warm house is comfortable and livable; a cold house isn't. Cars run and planes fly because of controlled fire. Fireplaces and campfires are good for souls and toes. Fire is necessary and important; we could not outlaw it if we tried.

Restraint is the difference between a sizzling steak and a gutted house. Show the child both and show them the difference. By showing them, I mean demonstrate safely the message. Don't just deliver a sermon and walk away thinking that the job is done.

You might show them a house in the neighborhood that has burned down recently. There's bound to be one; there always is.

Show them how you cook meat. When it is done, let them touch the meat. Children make their own connections.

You know your children and what will make the point, and they'll let you know when they've got the point.

The problem of children and matches does not extend only to those cases where the whole house, with everyone in it, was leveled by fire. That's bad, but less extreme cases are very bad, too. A nine-year-old boy was playing with matches with his eight-year-old sister. He set the whole matchbook on fire and burned his hand and dropped the matches into the lap of his sister's dress. Every part of her skin that was covered by the dress was completely destroyed. He had first-degree burns on his fingers, but she died after six days of agony. Where there are children in the home, matches should be treated with the same respect as poisons.

Children emulate their mothers and fathers, to the point where they play house and set up little tables to eat and serve each other. Naturally they don't use the stove, so the food is make-believe or consists of cookies, milk, or soda. If Mommy and Daddy eat by candlelight once in a while, the kids will want to do that too. If Mommy and Daddy smoke, the children probably will at least go through the motions of lighting up. You can't police your children twenty-four hours a day, but you must keep the fire at a make-believe level. No real matches.

Totally discouraging the children from make-believe would be hypocrisy. They're just trying to imitate you, to be like you in ways that are important to them. Obviously, it is also important to you that they imitate you.

Often the parent lights the cigarette and allows the child the privilege of blowing out the match—Is this luring, tempting, egging him on closer to his own real matches?—a very real danger. If you've gone this far, to give him a whack when he tries to do it next time would be just cruel. He couldn't understand it. . . . You're the one who's luring him this close into the fantasy situation of a match and smoke.

If you find burned matches anywhere around the house and you don't know where they came from, you'd better suspect your children are playing with matches. And you had better do something about it. It is an issue you absolutely cannot duck. If, after you have confronted and explained and even punished your child, it happens again, your problem is greater. Obviously, the child will not do it again *in front of you.* It will happen in the cellar, where the gasoline shouldn't be. It will happen in the attic with all that dry wood and dry stored things. It will be in the bedroom, the closet, behind the couch. It will be when the baby sitter is there. It will be when you went out and left him alone "for just a minute."

My most difficult job, in talking to the parent, is persuading him that the problem will not just go away if he ignores it. Far too often this confrontation is after the fire.

In the Baltimore Fire Department, we are fortunate to have with us a gifted young man named Lieutenant Tom Herz. He is our "Burn Injury Officer." Several years ago, he was severely burned while fighting a fire. He spent over a year in the hospital recovering from the loss of his hands, ear, and other painful scars. He's pretty effective talking to some of these kids. He has the commitment to get the message across.

If your problem is serious, call in the local fire prevention man, a uniformed fireman, and have him deal with the child. Sometimes the "official" uniform will make a lasting impression, especially if the fireman tries to win the child over, instead of scaring hell out of him.

If that doesn't work, it is definitely a job for the professionals. And I don't mean fire fighters.

There comes a point where these "innocent arsonists" are not so innocent. They know fire will burn them if they get too close and it will burn anyone or anything else that gets in its way. That doesn't seem to deter them. They know people burn; they're not afraid of other people's scars. It makes very little of the desired impression. These are not normal children at the threshold of understanding.

My experience with children is that they usually realize in some way when they have done something wrong. The kid who has set a fire downstairs comes running upstairs, goes to his room, and doesn't tell his mother that there is a fire.

Children learn from other children. There have been situations where neither parent smoked; the child got matches by going to the drugstore with a penny and saying "My mother wants a pack of matches"; or he can find them on the street, get them from a friend, get them from cigarette machines, steal them. Nobody's going to put him in jail for stealing a book of matches worth half a penny. So with a recurrent, determined problem like this, it doesn't boil down to hiding *your* matches. The world is full of matches.

If you have this problem in your family, do not—repeat, do not—leave that child unattended. Do not go out and leave him alone in the house. If you use a baby sitter, tell her or him what the problem

J.S.

is, alert her thoroughly in your fire plan. Be sure in your own mind that that baby sitter can really handle the problem while you are gone.

There was a child in our fourth-battalion area. The baby sitter ran to the store. The child was fascinated by the coal stove, so she got the door open and shoveled some coals out, took them across the room, lifted the cushion up, and dumped them in the overstuffed chair. When the baby sitter got back, four fire fighters were trying to breathe life back into four children. Two were dead on arrival and one died the next day. The child who set the fire survived.

If you detect a problem with a neighbor's child, tell the parents and insist that they get help. Keep that child out of your house, keep your children out of that neighbor's house, and stop your children from playing with him until this is resolved. If the danger continues, you are playing Russian roulette not only with yourself but with that child, the rest of the family, his playmates, and the neighbors. Who knows how one of his fires will end?

There may come a point that the child must be put somewhere where he or she can't light fires. And that is the best available solution until that child *won't* light fires; or play with matches, as some call it euphemistically; or commit arson, as others call it. Arson is defined by intent and not the age of the arsonist.

Juvenile pyrotechnics aside, no children should ever be left alone or unsupervised. So now it is fitting to discuss baby sitters, their selection, and their responsibilities. If your children cannot be left in the care of a competent mature person, they should not be left at all. Take them with you or don't go. That may be heavy, but the responsibility of being a parent should never be taken lightly at any time.

One in every twenty residential fires is directly attributed to actions of young children. In most of these cases, the absence of parent *and* baby sitter did not involve an unexpected emergency or economic necessity. Far too often the absence was due to rapping with a neighbor or downing a few at the friendly neighborhood tavern. In my thinking, under the circumstances both are criminal on the human level and legally may open either of them to the charge of being an accessory before the fact to arson or manslaughter.

You should know your baby sitter before you hire her or him. Check into the baby sitter's training, character, and family background. Have a preliminary interview (or several) with her before you entrust your children to her. You must trust her sense of responsibility and her liking for children. It is best to use the sitter who meets all these requirements regularly and it would be best if she lives nearby and knows your neighborhood.

Before leaving, acquaint the sitter with the children and pets, especially the watchdog. Explain your instructions to her verbally. Put the most important ones in writing, including phone numbers where you can be reached and those of the nearest neighbor or relative, the fire department, police, family doctor, and hospital. If there is no phone, explain what she must do.

Be sure the sitter understands—really understands—your fire plan, and be assured that she will follow it to the letter:

1. Escape out of the building with the children.
2. Account for all children.
3. Notify the fire department.
4. Then she is to notify you.

Carefully instruct her in your household fire precautions and caution her particularly against allowing the children to play with matches, or anything electric. If it is past dinnertime, the children should be kept out of the kitchen area. If you have a problem child, be *absolutely certain* that you fully explain to the baby sitter what the problem is and what she is to be especially alert for.

If the sitter is to bathe the youngsters, be sure she has such experience; if not, be sure you instruct her thoroughly. Leave, and explain the use of, a first-aid kit.

Be sure she understands that any fire is an immediate emergency evacuation situation. Her first duty is to evacuate the children and not to fight the fire. The children's lives are her responsibility and nothing else. If she smokes, be sure to leave her safe ashtrays and drop some hints.

Tell the sitter when you expect to be home. If you can't make it on time, be sure to let her know.

If at any point, your baby sitter warns you of a dangerous household hazard—fire or otherwise—stop, listen, and look, then remedy the situation.

To the baby sitter:

Be sure you really understand all instructions. Ask questions until you do.

Stay with the children before they go to bed.

Look in on them often once they are in bed.

This is your job, not a social occasion. Your friends are not to visit you while you're working unless it is known and completely approved of by your employer.

Do not use the telephone for personal calls; you may block an important incoming call.

Keep the television and the music level down; you are being paid to hear and respond to any sounds of distress. You must be able to hear the children if they cry or are having trouble.

Remember, if you are the sitter you are in charge; don't let the kids give you the "Mommy said it's okay" jazz. Be alert and be firm.

CHAPTER 18

Smoking and Matches

More than half of fatal residential fires each year are set by smokers and smoking. They are caused by cigarettes, cigars, and pipes—or matches or ashtrays—that smokers thought were out.

Almost all of these fires—about 90 per cent—ignite when the smoking material connects with bedding or with an upholstered sofa or chair. The average mattress or overstuffed-chair fire produces more toxic smoke than flame. Although the fire does not usually spread far, it can, and it takes a very heavy life toll.

Most of these fires are at night, caused by someone falling asleep while smoking, or by someone emptying a hot ashtray into a wastebasket before retiring, or by someone "forgetting" a lighted cigarette —someone merely tired or perhaps impaired from a little alcohol, or a lot. Or, possibly senile. A lot of older people set themselves and their dwellings on fire this way every year.

I personally know of a situation where the family of an elderly man, who was only able to move his hands, put a piece of emery paper on his wheelchair. He was allowed to strike wooden kitchen matches on it so he could smoke. Sooner or later it had to happen: The head of a match broke off in his lap, and he was cremated.

Do not give matches or lighted smokes to anyone whose judgment, perception, and physical ability are impaired. This includes the senile as well as the mentally and physically handicapped. Fires caused by this kind of situation take a horrible toll among persons over sixty-five. Some of them may be well able to take care of themselves; they are independent and self-sufficient, but they are not quite so alert or perceptive or quick as they used to be. If you really love that person, help him or her and the rest of your family avoid that first mistake. If that person loves you, he will understand and will not take offense if you light his cigarette for him and sit and talk with him while he smokes it.

If you are that person, think seriously and deeply about the responsibility you have to yourself and your loved ones and your neighbors; think to the depth of your soul about that terrible potential for pain and death and destruction you hold between just two fingers. And consider how just a little pride could save a lot of lives. Fire is not selective and hardly ever touches just one life. I have nothing to say against smoking tobacco, only against smoking houses.

We have an enormous problem with mattress fires. I would add here that, although I recommend the use of one smoke detector located between a family asleep and the rest of their residence, I would strongly advise that if anyone smokes in a bedroom, that bedroom also should have its own smoke detector.

Obviously no one should smoke in bed, but equally obviously a lot of people do and a lot of these people get burned or killed from that mattress fire. And sometimes other people in the home get burned and killed when this fire sets the whole damned place on fire.

There will be some heavy resistance to what I am going to say now. If you have a fire, no matter how small, and you put it out, no matter how thoroughly, you must call the fire department and say you have what you think is an extinguished fire and ask them to please send somebody by to check it out.

Now here's where some people start to panic. Their sofa was smoldering or their bed. They discover it early in the game and carefully pour a glass or two of water on it and it goes out. There are a lot of things tugging at your mind to make that call. . . .

You want to go back to bed.

You're sure the fire is out; it was such a little one anyway.

You don't want the attention.

You don't want the neighbors to see any kind of uniform coming to your place.

A fireman in uniform has that military look; he looks an awful lot like a policeman . . . which is good and is not so good . . . for your purposes. You feel that the fireman who comes to check out the fire that you "put out" is going to look at everything in the place and you are not sure you want your privacy to be violated even a little bit. Some people experience a certain paranoia at this decision.

Well, let me tell you something about mattresses. They are in a category all by themselves. Of all fires "extinguished" by occupants, this is the most treacherous one. About as safe as a sleeping rattlesnake.

Don't smoke in bed. Don't smoke when your judgment is poor, whether from fatigue, alcohol, medication, or other drugs.

If you're compulsively tidy about ashtrays, remember what a mess it is every time your kid spills just one glass of water. Imagine a pumper delivering a hundred gallons of water into your living room. We have to do it in order to put out that fire started by emptying that ashtray with one hot cigarette into a wastebasket half full of paper. Maybe five hundred gallons, maybe a thousand. And then think about the fire damage.

I remember one twenty-nine-cent wastebasket that happened to. Lady dumped an ashtray and then went to bed. There was very little fire damage, but she had sooty stuff like someone had blown carbon cobwebs all over the house, or run a bucket of tar through a cotton-candy machine. It was a hell of a mess.

Dispose of everything burned as if it were still burning.

Dump the ashtray in the morning. Or flush the contents down the toilet. Don't leave the ashtray on combustibles. Make sure all butts are *in* the ashtray, not resting on the edge. Handle that ashtray and those cigarettes as if they were a bomb—they are. One cigarette plus one ashtray plus one wastebasket plus one gallon of gasoline equals sixty-six sticks of dynamite if they come together. If you take away the gasoline, you have one destroyed house and a few dead people instead of a whole block of devastation. It's still just one cigarette anyway you look at it. It's still fire.

Before you hit the hay, think about hay burning and check out all upholstered furniture for heat or smoldering.

Recently we had a house fire that killed six people. There seemed to be some discrepancies in the principal witness's account of the fire, so we did a full arson investigation. It definitely turned out not to be arson, but the person who accidentally set the fire will carry a terrible burden of guilt the rest of his life. He went to sleep on the sofa in the living room of a three-story house. He took out a cigarette, but didn't have any matches so he lit it on the kitchen gas stove and returned to the sofa, lay down, and fell asleep. He was awakened by the heat from the burning sofa. He got up and ran out the back door. The witness had been drinking, and he did not remember putting the cigarette out.

The fire originated in the first-floor living room, completely consuming a sofa and chairs in the room. The fire spread up the walls, across the ceiling, out the door of the room, and up the stairway to the second floor. The fire then traveled down the hallway to all the rooms and up the third-floor stairway to the third-floor hallway into all the rooms there.

The electrical fuse box was examined and showed no signs of having been overloaded. There were no electrical receptacles or wiring in the origin of the fire, nor were there any heat devices. The oil-fired furnace in the basement was examined and showed no signs of having caused the fire. All heat-supply pipes and doors on the furnaces were in their original positions and showed no signs of any explosion or malfunction. There were no signs of any fire around the furnace, furnace motor, or the oil-supply tank in the basement. The ceiling above the furnace showed no signs of any fire.

There were metal registers in the floors that supplied heat from the furnace, and these showed no signs of fire. No signs of any flammable liquid could be found.

The rapidity and ferocity of the fire were caused by the open stairway and a delayed alarm by the occupants. The fire apparently was in the smoldering stage for a considerable length of time, causing superheated air and smoke to rise to the upper floors. When the first-floor rear door was opened, it permitted oxygen to enter. That contributed to the complete combustion of all furniture, wood trim, wooden stairs, wallpaper, and other combustibles on the first, second, and third floors.

The springs of the sofa were severely annealed indicating a slow, smoldering type of fire, the kind that produces a heavy concentration of carbon monoxide. The toxicologist added that the high concentration of carbon monoxide in the victims' lungs indicated prolonged breathing of it; it also indicated that all were alive before the fire.

The Medical Examiner's Office stated that its examination revealed that all the deaths were caused by smoke inhalation and carbon monoxide poisoning, adding that there were no signs of any violence on their bodies. Note: The high concentration of carbon monoxide in the bodies of victims was indicative of a slow, smoldering type of fire. The lower concentration of carbon monoxide in the body of the little girl can be explained by her age (one year) and breathing capacity. The lower concentration of carbon monoxide in the body of the twenty-three-year-old male can be explained by the fact that he at one time had left the building and re-entered it.

The principal witness is alive and . . . well? One cigarette out of control and six people never knew what happened.

CHAPTER 19

Heating and Cooking

Nearly one third of all residential fires and residential fire deaths originate in heating and cooking, by malfunctions and accidents. The electric or gas kitchen range and oven, the oil-fired heating plant, and the gas-fired water heater are the largest culprits by sheer weight of numbers. Other fuels used for cooking and heating—wood, coal, kerosene, gasoline, and charcoal—are exceptionally dangerous but are used less than formerly.

There is a basic distinction between cooking and heating fire sources. Cooking is out in the open and localized in the kitchen. Heating is characterized by concealment and its weakest link—its danger points—may lie hidden anywhere in or beneath the dwelling.

Fire is energy, wild and destructive energy, when unchanneled. Cooking and heating are domesticated energy uses. The tame, useful flow of this energy throughout any house is more or less constant, but if there is any weakness or imperfection in the energy-restraining system, sometimes there is a breaking point or rupture during cold-weather overloads.

If you have any questions about any wiring, any appliance, or any suspicion of any gas or fuel leak, do not hesitate to call the Gas and Electric Company, the fire department, or a professional repairman—pronto! Do not attempt to repair it yourself. Do not look for a gas leak with a lighted match or a short circuit with your fingers.

(You think nobody looks for a gas leak with a match? I quote the following incident:

"Yakima, Wash., Nov. 12, 1957, $5,000—One unit of a 7-unit motel was destroyed by a gas explosion that occurred while the serviceman for the Gas Company was on the premises looking for a gas leak with a lighted match. The serviceman suffered nonfatal burns.")

Let's talk about cooking first.

Don't store any flammables near the stove, for obvious reasons.

Don't cook anything on the stove that the stove isn't for. One man thought his aerosol can of lacquer was too cold and warmed it on his kitchen range. It went off like a rocket and stuck right up there in his ceiling.

Control the stove. In most houses, the kitchen is the "real" living room, and lots of folks hang out there a lot because it is comfortable and friendly. For some reason, this is where people usually end up during a party. There's a lot of social activity in most kitchens, so you have to make sure nobody catches on fire while dinner's on the stove. Be particularly careful about children in the kitchen. There's fire here and kids need the same supervision and instruction here as they do with matches.

The proper placement of pots and pans on the stove is important. Use the back burners as often as possible because children cannot reach them as easily. Limit and control the front burners especially, because children have the habit of going into the kitchen to see what is on for dinner. It is natural and automatic to lift lids and look in the pots and pans. But a youngster of five or six, who can't see, may reach for the pot and whatever is cooking in the pot will be cooking all over the child. Or he can set his sleeves on fire reaching for it, especially over a hot front burner. And so can you or any other adult if you have a lapse in your carefulness.

When you're cooking, it might be wise to roll up your sleeves or wear tight-fitting garments—not the loose, dangling, or ruffled-type sleeves. Clothes burn. And put that cookie jar somewhere away from the stove, somewhere the kids can reach it without getting burned.

Some wisdom has been built into the products of some manufacturers, who short-cord certain kitchen appliances such as electric coffee pots, waffle irons, and toasters. They used to have five- or six-foot extension cords, but the manufacturers have caught on to the fact that they ended up hanging over the counters. Kids were coming through the kitchen and grabbing the cords and pulling the things down on their heads. So they've shortened them quite a bit. If any cord in your kitchen is too long, an electrician or an electrical repair shop can shorten it for you.

Sooner or later, everybody has a pot of food catch fire on the stove or in the oven. A frying pan with grease in it is probably the most dangerous. It can spatter and it can spread and catch a kitchen cabinet or curtain or your clothes. Of course, don't let it happen; but if it does, what do you do about it?

Fire extinguishment is the exact reverse of fire ignition. Fire is fuel, air, and ignition temperature. Remove the ignition temperature first. Turn the heat off, both range top and oven. One fire is enough. Do not remove the fuel—the flaming frying pan. Do not attempt to carry that hot pan anywhere. Suppose you got halfway up the hall with it, couldn't hold on any longer, dropped a blazing pan of burning grease all over your rug or down a stairway or all over yourself—that would be it! Leave it be.

Now remove the oxygen. Put a lid on the pan—the tighter the better; close the oven.

Bicarbonate of soda—ordinary baking soda—will put out many *small* fires by interfering with combustion, in part by lowering the heat but primarily by creating a chemical barrier that keeps the oxygen out. Baking soda is a white powder that usually comes in a rectangular cardboard box. Do not confuse it with baking powder, which usually comes in a round box or can and looks like flour. *Bicarbonate of soda* can put the fire out; that's the one you want. Corn starch and flour can cause a dust explosion.

Do not use water on a grease or electrical fire unless you want to set the whole kitchen on fire with you in it. Water will spatter that hot grease from wall to wall in a flash. If the fire is electrical, in a toaster or electric range, you will not put out the fire, and you may electrocute yourself. Disconnect the appliance safely, *if you can.* If you have doubts, call the fire department, alert the others, and get out.

There was an instance a few years back where a young woman was emptying some trash. It contained flour. She dumped it down an incinerator shaft. When that flour hit the bottom of that incinerator, it blew everything back up again and blew back into her face. Flour should not be stored above a stove.

Remember, whenever you have a fire in your house, regardless of how well you have extinguished it, you should always call the fire department for a check-out.

It's free and it could save your life.

There are lots of odds and ends in the kitchen that can cause fires. Here are some do's and don'ts.

Don't heat sealed cans or jars on the stove. Open them and cook them in a pan, or at least open them and heat them in a pan of water.

Don't dry clothes or anything in the oven or on the stove.

Don't let grease accumulate anywhere in or near the stove.

Don't ever cook with charcoal in the house, even under a vented hood or in a fireplace. The invisible smoke from charcoal is almost pure carbon monoxide and it is lethal. The only place for charcoaling is outside in open air. Incidentally, a damp bag of charcoal briquettes can ignite itself; so watch how you store it. So can damp rags or paper. Keep it all dry.

Watch what you throw into the wastebasket. Grease on paper trash is ten times as dangerous as paper trash. Watch your smoking and your cigarettes.

Wooden kitchen matches are dangerous. They flash and shoot. Rats and mice can strike them by chewing on them. I would say they really have no place in the home; but it's your home. Be very careful using and storing them.

Don't leave the potholder on the stove and don't store it too close to the stove or above it. The same goes for paper trash, or even plastics. The empty waxed-cardboard milk container is dangerous.

If your stove has a hood vent or exhaust fan, keep it and the filters especially clean of grease buildup. Filters can catch fire; so can grease even high up in a hood or in an exhaust duct. Grease buildup on exhaust fans can overheat the fan motor and it can catch on fire and ignite the grease.

No gasoline, cleaning fluid, or flammable liquids, gases, aerosols, should be kept in the kitchen at all—ever. The kitchen is full of pilot lights and ignition sources. A gas range top has one. The exhaust fan motor has an electric spark. The refrigerator motor spark is ever present, as is that of the dishwasher and any other running motor. Even the toaster is hot enough to ignite volatile vapors and fumes. Keep your nose open for gas leaks.

Don't use any electrical cord that heats up or is frayed or worn; if you hear any small appliance making a funny noise or if it's overheating, sparking, or smoking, don't use it. Unplug it and get it fixed before you use it again.

Some kitchens have wood- or coal-burning ranges. Priming or kindling them with gasoline, kerosene, or lighter fluid is playing Russian roulette. The kiss of death is a big, hot "whoosh!" and that could be it, right then and there. These fumes are dangerous.

Gasoline camping stoves are bad enough in campers. They have no place in the home, and they must be emptied outside before being stored inside. Kerosene hot plates really don't belong inside. There are easier, safer cooking units on the market.

Electric hot plates should be used with extreme caution. They draw heavily on electric wiring and can short it out. They get very hot and should be used only on an insulated, noncombustible surface far from flammables and combustibles. No bedroom should double as a kitchen; keep hot plates out of the bedrooms. Don't get paranoid but be reminded that many people die each year because these things do happen. Your kitchen is as safe as you let it be.

Smoldering mattresses aren't the only fires that can kill you without actually burning you. Recently we arrived too late to help a man who forgot to turn off the flame under a pan of grease on the stove. His nine-year-old daughter came home from school and took a nap. The man decided to take one too. At the opposite ends of the house they died in their sleep without waking. There was no fire outside the frying pan. The smoke got them.

All heating plants, furnaces, water heaters, and any heavy gas or electric appliances should be regularly serviced by professionals.

Don't store flammables or combustibles near any heating units or any heating duct or radiators.

Don't block or clog any ducts or registers.

Don't dry clothes on any heating units, ducts, or registers.

Keep your thermostat—water heater and heating plant—as low as you can be comfortable with. I'm not saying that to conserve energy; I'm saying don't overload them or burn them out by demanding more of them than you really need or than they can safely do.

Have any fuel leaks checked professionally immediately.

As it happens, the basement, or garage, where most heating plants and water heaters are located, is exactly the place most people store their most flammable items: gasoline, kerosene, paint, varnish, varnoline, paint thinner, brush cleaners, alcohol, solvents, cleaning fluids, lacquer, and lacquer thinners. Need I say more?

At the first signal of any problem with your heating plant, turn it off and check it out. Call a professional. Have you ever heard of oil-burner backdraft? It's an explosion, caused by delayed ignition and the buildup of a wrong ratio between fuel and air. The ignition spark fails a couple of times, but the atomized oil keeps spraying and creates more vaporization than is needed. Then the ignition spark connects and "whoosh!" the thing explodes out into the cellar. Normally, that in itself is not too bad; it's trying to tell you it needs some professional attention. But if there are flammables nearby, they can go right up.

When you go on vacation, remember all the pilot lights in your house. We have learned to take them so much for granted that we don't think to put them all out. If nobody is in that house for two weeks, the least you can do is leave a window or two open a small amount, top or bottom, to ventilate the house. And make sure the storm windows are vented or opened slightly too. (Some of them have venting manufactured into them.)

Pilot lights *can* go out. A burp in the gas supply or even a leak in the water heater could extinguish the pilot light. As a rule, pilot lights don't have pilot lights. When they go out, they stay out until someone relights them. Some of them have shut-off safeguards against that, but you'd better find out. If you smell gas, report it. Most utility companies will check it out free. Even if they charge, it's worth the price.

CHAPTER 20

Electrical Hazards

Electricity is compressed fire in a wire. That wire can travel everywhere. As long as it doesn't leak or get compressed too much, it remains fire in theory only. But electricity itself can kill you; it can burn or zap you. It can ignite flammables and combustibles by high temperature or by spark. Ubiquitous as air and fuel, it is indeed instant fire.

One in six household fires is electrically ignited. Two thirds of those originate from electrical wiring; one third begins in electrical equipment: appliances.

There are often symptoms of electrical distress before an actual fire is started. Occasionally those symptoms are gradual and progressive, often they are sudden and assertive; sometimes the fire is the only warning of electrical trouble. Except for extension cords, wiring is usually hidden; even in appliances, the circuitry is well out of sight.

So you must be alert to external symptoms your electrical systems may emit as signals to your senses: fuses blowing; circuit breaker tripping; lights flickering; heat in any wire, plug, or receptacle; excess heat in any appliance; smoke from any wiring, receptacle, or connection; sparks or arcing from any electric source; the characteristic odor of electrical distress; shocks when appliances or wiring is touched; the sound of snapping, crackling, popping, or sparking; the sight of broken, frayed, or worn insulation; the sight of bare wire; the hiccuping of thermostats, controls, and switches; the less than 100 per cent or irregular, operation of any control or appliance.

The very first rule is: Do-it-yourself electrical installation or wiring should *never* be undertaken by an amateur, only by a trained and licensed professional electrician. You may be an excellent carpenter or antique restorer or champion mechanic, but lion tamers aren't snake charmers. Jacks-of-all-trades are not masters of this one. Everybody is doing it himself and half of them are doing it wrong. Ditto plumbing, where gas lines and pilot lights are involved: This is one area where the do-it-yourself handyman must not trespass on professional territory.

The second rule is: At the very first symptom of electrical distress, turn the power off. Turn the switch off, pull the plug out, or disconnect at the source the fuse box or circuit-breaker panel. Then have an electrician make repairs before you turn it on again. The problem will not go away or get better if you simply let it cool down for a few hours; it can only get worse.

Because electricity is normally out of sight in a household, it is doubly deceptive and dangerous. You can't ignore or wish your problem away because it is abstract and hidden, any more than you can will away a gas leak.

If you have an electric fire, call the fire department immediately. Disconnect or unplug the appliance or wiring at the plug, the switch, and the panel. Don't touch anything. Do not put water on a live electrical fire; it could electrocute you. Water can put the fire out only *after* the circuit is disconnected, but CO_2 or dry chemical extinguisher will do better. At any stage of involvement with an electrical fire, call the fire department. It's not a law, but it should be.

Your fuse box or circuit breaker is supposed to turn the current off when a circuit becomes overheated or overloaded or if it short-circuits. That is why it's there and that is its only job. If you put pennies in the fuse box, you might just as well take a match and start the fire yourself. That way you have the advantage of knowing when the fire will start, instead of having it erupt days or weeks (or hours) later while you are sleeping. The number of fires started by short-cutting this life-saver beggars the imagination.

No appliance or electrical equipment—even the lowly extension cord—should be without a seal of inspection and approval. If it isn't certified, don't buy it. If it's a gift, don't use it. If you can't bear to throw it away, don't pass it on to someone else; that isn't friendly. Get a qualified electrician or electrical appliance repair shop to inspect it *thoroughly* and take his word to chuck it if he says it's dangerous. Just throw it away. If any appliance is not doing the job it's supposed to do, get it checked out promptly. Underwriters' Laboratory and other agencies certify *new* appliances and devices; when these things get old they need reinspection at a professional level. Anything can deteriorate; a UL certification doesn't mean too much after ten or twenty years.

Cords get frayed or brittle and crack with age. I've never seen an iron yet that didn't have a cord that eventually frayed if it ironed enough shirts. So watch these very closely; they tend to deteriorate. Don't keep using that iron just because it keeps on working if the cord to it keeps getting hot everytime you use it. That is just plain dumb.

A woman called in a fire. Her electric iron burned through an ironing board and set fire to the room. She wanted us to say the iron was defective, for insurance purposes I assume. That iron was very effective; the woman was defective. We had no way of knowing whether she had turned it off or not. She admittedly did not unplug it. Anytime you leave an iron, don't just turn it off; turn off the switch *and* unplug it. There are other ways irons can set fires. There was the case of a woman who had a spot on her dress and tried to remove the spot without removing the dress. I think she used alcohol. She put the dress on the ironing board—still wearing it—and tried to dry it with the iron. The dress caught fire on her. She ran from one room to the next, fanning the flame. She died about an hour later.

Under no circumstances, unless you are professionally qualified, should you be your own gas maintenance man or your own electrical maintenance man. You don't know what you are doing and you are going to end up with a fatality. There are too many variables where electricity is concerned. The glues used in plywood and veneers are flammable; remember that, when you are redoing your clubroom and messing around with wiring, laying carpeting, and putting up new walls and ceilings. It may look bright and new and shiny, but you may have built a death trap into your basement by some of the materials and combinations of materials that you have used.

The cheapest is not the safest; sometimes it is not safe at all. I'm thinking of the average do-it-yourself putting up paneling at $2.99 the sheet around the basement, burying some wiring at the same time, and then covering up some ductwork or an old electric socket he didn't know what to do with. He was wise in thinking not to mess with it; but in actual fact he *has* messed with it and when the spark comes from behind the paneling, he doesn't know what has happened and he has big trouble on his hands.

One man went so far with his do-it-yourself that he rewired his whole house. He burned the entire second floor out and the roof came competely off the building. We had to put it down to get the fire out. He had a regular electrical octopus strung out all over the house. In one junction box, with four outlets, he had four more coming off each one of these—that adds up to sixteen.

Here's a quick rundown on specifics.

Special wiring is required for air conditioners, electric ranges, and heaters because they draw such a heavy load of current; for this reason, they need separate circuits and their own breakers. They must all be properly grounded, and their circuits should not be overloaded. All built-in ranges and ovens should be easily disconnectable from all ungrounded connectors. Air conditioners, fans, and refrigerators can overheat because of dirt or dust clogging their motors or filters. Keep them clean. In refrigerators the same applies to the refrigerant coil; have it cleaned and serviced regularly.

Electric irons should be approved and have approved temperature-limiting devices; each should have an approved stand and have an approved cord in good shape.

Electric clothes washers and dryers should be well grounded and have some means of quick disconnect. Lint in dryers is a high-risk situation; lint filters should be cleaned before and after each use and the machine regularly checked out for lint deposits. Clothes can catch on fire in unsafe machines. Don't leave the machine running unattended. Read all the small print about synthetic fabrics, plastic, rubber, and foam. Some things have no place in the clothes dryer. One couple, in a laundromat, had some clothes scorch in the dryer and start to smolder. They threw away several ruined garments and took the rest home. The clothes basket set fire to their bedroom.

Remember not to dry clothes on or in any other appliance. Fires have started from drying clothes overnight on a hot-air register, in a low oven, or in front of an open fireplace.

Electric blankets and heating pads are hazardous if not disconnected after use. An electric blanket, rumpled up with the other bedclothes, can set the bed on fire. Believe it or not, a water bed can start a fire if it is drained and its heating element not disconnected.

Lamps and light fixtures should be used with some caution. Incandescent lights generate heat and should have plenty of cooling by ventilation and not be in contact with flammables or combustibles. Fluorescent tubes are relatively cool, but their ballasts can get very hot. Replace defective or worn cords.

Don't run cords under rugs, doors, or windows; don't allow them to be otherwise mashed, kinked, squeezed, or rubbed. Keep furniture clear of outlets so its weight doesn't rest on cords or press against plugs in outlets.

A television set should never be placed on a bed where it can sink down into the bedding causing heat buildup. TV sets, and radios, require ventilation to cool them. They should never be recessed or built into any wall or bookcase without adequate air circulation designed and built into the installation. Don't leave a TV set on when no one is watching it; and if you are sleepy and alone, turn it off. Make sure it has plenty of clearance from draperies, curtains, and bedclothes.

Remember that instant-on TV sets are always on. That means another ignition source in the house and it is another source of heat accumulation and possible malfunction. If you are going away on vacation, unplug the set.

The bathroom is no place for TV. To be a purist, it is no place for anything electrical while you're wet. Never, never touch anything electrical from the tub or shower.

Outdoor antennas and lead-in conductors for home television and radio equipment must be securely supported, be of corrosion-resistant material, and be kept clear of electric light or power circuits. Masts and other metal structures supporting antennas should be protected with an approved lightning arrester.

Portable heaters should be so designed that accidental tipping or upsetting cannot cause a fire; they should be provided with an approved heat-resistant cord.

Christmas trees, decorations, wrappings . . . beware! Do not use any decoration that has not been flame-proofed nor any electrical ornaments or lighting that has not been UL approved. Do not use electrical fixtures on metallic trees. Need I say more?

CHAPTER 21

Sparks and Stuff

Open fireplaces and coal and wood-burning stoves once used to be the only heating and cooking systems in early America. Now they are largely supplemental and warm the spirit far more than the house. In actual fact, they tend to act as a heat pump, drawing inside heat up through the chimney to the outside by the convection of their thermal currents.

The installation of the acorn stove, the Franklin stove reproduction, or the permanent in-wall installation is no job for a novice. All piping, ducts, and connections must be insulated and fire-safe. As a rule, the chimneys cost more than the basic fireplace units and competent installation may run even more than the cost of all materials. A unit costing one hundred dollars could easily cost a thousand, installed; ten thousand, if improperly installed or used. Deal with a professional and talk with some of his satisfied customers. This is *not* a job for do-it-yourselfers.

If you have a working fireplace, have it inspected periodically: flue, damper, chimney, firebox unit, and connections. The chimney should be cleaned once a year to remove dirt, soot, unburned fuel—and bird nests. Don't let anything block your flue—top or bottom—and don't, yourself, block your flue with rags or any combustibles to keep out the cold draft. If your damper doesn't work, get it fixed. In some localities, your fire department will furnish a fireplace inspection. If not, the same contractor who installs units can do it for you. Losing your house to a fireplace accident is like making it all the way through a balanced meal that is really good for you and choking to death on the dessert. If smoke leaks from your chimney, fire and flames can, too.

Overfiring can ignite adjacent combustibles anywhere on the fire path to the outside. Ceiling and roof joists can ignite from heat alone. Combustibles too near the fireplace or the chimney, even in the attic, can catch fire. Sparks can ignite the roof or nearby grass, brush, or rubbish. The roaring fire should stay within the system.

A backdraft from weather conditions can flood your house with smoke or blow some of the fire out onto your rug. Nothing highly combustible should be near the fireplace. Furniture is also combustible. Watch out for draperies and the Sunday paper.

Always use a fire screen. Avoid any type of wood that shoots, sparks, or explodes. Use dry wood. Coal isn't a good idea because of the fumes. Don't burn cardboard or household trash in the fireplace. You could set your roof on fire or generate poisonous smoke.

Never start a wood fire with kerosene, gasoline, or lighter fluid. Use paper and kindling to start the logs; it's more fun anyway. Never goose up a fire with these fluids either.

No charcoal in the house!

The bottom of a wick in a candle comes close to touching your house. Your candle belongs in a fire-proof base; it should not be near any combustibles or flammables (and that includes long flowing gowns and sleeves of lounging apparel). If used while dining—and this includes candles warming coffee pots and fondue sets—it should never be left unattended. You are going to insist I'm unromantic when I tell you a candle has no place in your bedroom.

Your hair is flammable, especially with lacquer spray net on it, so don't go up in flames. Even if your hair flashed and didn't really hurt you, you nevertheless could be startled into dropping or throwing the candle—anywhere.

Fireworks are . . . Finish that statement in one word and win one or more of many prizes ranging from your finger or your eye to your house or your life.

Open burning of trash or leaves in fields and open lots is prohibited in many areas because of conservation and pollution awareness. Because of its danger in the hands of amateurs, it should be. Trash and grass fires are a hazard to your house from the outside in, as any torch is. A storekeeper was burning trash in a barrel just outside his downtown store. It got away from him and went into the store, gutting it, and burned two children in an apartment upstairs.

You're paying the trash collectors. Let them do your job.

Mow that vacant lot. Rake it and bag the rakings for the trash man. Fallen leaves near buildings are especially dangerous.

In rural areas, controlled open burning is sometimes sanctioned by the local fire department; often, such burning must be done in the presence of fire fighters. Please keep it that way. Look at California and see what grass fires can do.

Today's treasure is tomorrow's trash. If any article has outlived its usefulness, be sure you outlive it, too; get rid of it. It has no place in your house if you don't need it and will never use it.

Clean up your act. Trash and debris—boxes, paper, old clothes—under and on stairs, in attics, closets, cellars, and garages are pure fuel. They invite fire, block escape, and sometimes even flare up spontaneously.

Dumpster fires can be dangerous if they are close enough to a building or a car. The smoke they generate is wicked. But they are dangerous also in another way: A dwelling could catch while firemen are out suppressing a dumpster fire some wise kid lit for the pure hell of it. This kind of vandalism is on the level of false alarms and high-rise trash-chute fires. It's not funny. You won't think so either; in fact you'll get hotter by the minute just thinking about it as you wait for the fire department to come save you and your house.

Fires can begin from seemingly unlikely, out-of-the-ordinary happenings. Here are some examples.

One occupant of an apartment had a roach and bed bug problem and decided to fumigate his house himself. Sulfur candles are against our fire code, but some places still sell them. The instructions tell you to set them in a saucer of water, but, in this case, the guy set his in small metal-can lids with a little water in them. Then he lit them, closed the house completely, and left the premises.

The result was a complete burnout. We got a call from the next-door neighbors who thought all that smoke was coming from the sulfur candles until their own wall got hot and started smoking.

The water had evaporated from the lids and the heat burned through a table top, ignited the rest of the furniture, and the place burned down completely.

Sulfur candles are a menace; so is the general principle of leaving any fire unattended, whatever the application.

Another situation, having to do with unsupervised children: We received a call to a big, old apartment building. A kid had been curious to see what was inside a dumbwaiter. He rolled up a newspaper and made it into a torch so he could see. When it started to burn his hand, he dropped it, and it fell down into a lot of trash that was at the bottom of the thing. The fire rose to the top and went to the ceiling level of the highest floor and then traveled to the front of the building.

When the apparatus arrived, we saw a weird sight: Smoke and flames were coming out of the cornice and also coming out the first-floor level, but there was no fire on the second or third floor. That baffled us at first, but we figured it out quickly enough.

In another situation, an old man with a wood stove with an old chimney was getting smoked out because his stovepipe got clogged at the point it went into the chimney. He cut a hole in the partition next to the chimney, about three feet away, and stuck the stovepipe directly into the partition. Consequently, the smoke from the stove went into the partition, from there through all connected partitions and quickly throughout the entire building. Luckily, it was in the afternoon. You have never seen so many derelicts and winos, drunks and bums, cockroaches and rats abandon one building in such a wild scramble, in your life.

When we arrived, the smoke was pouring out of the roof eaves as fast as the inhabitants were pouring out the lower doors and windows. The blissful drunk who did this was blissfully dead from the smoke. There was no building fire, as such.

Willful ignorance and stubborn stupidity know no bounds.

There are two kinds of barbecues that can hurt you and your home.

One is the one for cooking meals outside. It has the same potential for igniting clothing as any other open fire, more so maybe because it is associated with fun, and fun tends to lower alertness.

The other is the floor register in the house which can act as a barbecue; if the lower part of the house or ductwork has flames in it, these can shoot straight up from a floor register and barbecue or block anyone attempting to pass over them.

Clothing is highly flammable. If your clothing catches fire, don't run or you'll cremate yourself by force feeding oxygen to the fire. Be still and either get out of those clothes pronto, or lie down and try to roll the fire out. If possible, smother the flames by rolling in a blanket or jacket or anything that will extinguish the flames. Apply water if you can or even shoot yourself with a fire extinguisher, but never in the face.

If your power ever goes out, be especially careful. Candles, the fireplace, and any makeshift torches can be dangerous, especially since your control over them is at an all-time low in the dark. Preplan:

J.S.

Have flashlights in your home and use them when you need them. Christmas trees and burning fireplaces or candles do not ever belong in the same room. Sparks are fire seeds that sprout like lightning.

If you'll bear with me, these will be my last words on gasoline:

There is probably no way to keep it out of the garage, because that is where the car, the motorcycle, and the lawnmower are. And I suppose this means you feel you have to store a spare fuel can of this—um—liquid. All I can say is, don't store more than you temporarily need. And don't use anything but a UL-approved heavy-duty gasoline safety can with a spring-type lid. Label it GASOLINE, so nobody plays with it or drinks it or does anything terrible like that. Never store gasoline in a plastic container; it can react or erode or puncture and leak. And never store it in glass, which can break. Keep it in the garage or an outside shed; never store it in the house, not even for a minute. Gasoline should never be stored in an area where there is an ignition point or a pilot light. I hope by now you know what they all are, including fireworks and light bulbs. And I know I'm talking you to death about gasoline. But that's preferable, don't you think, to the alternative.

Gasoline fumes can travel far and fast and, upon ignition, can travel sixteen feet per second back to the source. That's faster than you can run and slower than you can fly.

Watch it with the other flammables, too. Some of them are just as bad (almost).

If in doubt, read the label. If still in doubt, ask somebody who knows.

Some people are still cleaning things with gasoline and rags; some of them burn off paint or old tiles, ignoring what lies under that surface; the fumes of some of these things keep saying to them DON'T DO IT! But they do it anyway—and take their families and neighbors with them far too often.

I know an off-duty fire fighter—are you ready for this?—who was removing paint from his kitchen woodwork. He didn't want to use a paint burner because he knew it was too dangerous, so he used a flammable paint remover. A few beers later, the stuff was too slow for him and too messy. (It gets really icky, and the fumes can lay a bad high on you.) So he got tired of all this and perhaps influenced by the fumes he got out his paint burner, and he speeded up the process. Amen.

Lest I be accused of banging too long on the gasoline can and drawing sparks from the reader, let me move on to some exotic household explosives:

Remember I said that hair can catch fire or flash if exposed to open flame, particularly if the hair is lacquered with spray net. Imagine what could happen if you are humming away with a cigarette in your lips, teasing and spraying away with that aerosol bomb of hair lacquer. The more you tease it, the better the surface to oxygen ratio; and the more you spray, the more fuel you are adding—and you already have that ignition spark in the middle of your face.

You could go boom as your hair goes poof! Believe it or not you can take those spray nets and light the spray and remove paint with them like a Bunsen burner. Dynamite. People use these things every day of the year and smoke right through the operation.

Since it's probably illegal for you to burn any trash, perhaps you won't incinerate any aerosol *bombs*. Don't even consider it. Out for the trash man.

Another local man knocked an everyday, hardware-store propane tank off his cellar workbench and it sprung its top. When it started hissing, he ran outside. Good thing; the basement exploded in fire. It was only a little propane tank, about nine or ten inches tall. The pilot light from the water heater ignited the gas.

In the report on electrical hazards, I warned against electrical home installation and repairs. Now let me tell you briefly why only professionals should do gas installations and disconnects. I'm going to cite two examples:

Following an explosion, the father, the mother, and three children were killed in the ensuing fire. Two other children were injured. Investigation revealed a galvanized pipe from the propane storage tank had been fitted to a copper gas line in the house; electrolytic corrosion had eaten a hole through this supply line to the propane heater. The father had done a beautiful, neat job of his installation, but he neglected just one little thing.

In a large apartment complex, maintenance men installed a new gas range and oven in the kitchen and checked it out, pilot light and all, just prior to new occupancy. Then they locked up the apartment and left. Twenty minutes later, the whole place blew up. The previous occupant had disconnected and removed his own gas clothes dryer from another room and did not think to cap the live one-inch gas line that supplied *his* appliance.

CHAPTER 22

Arson Is Potential Murder

Professionals—fire officers, arson and insurance investigators, people whose job it is to sift evidence from ashes—believe that fully half of all fires of unknown origins are arson cases.

More and more we are coming to see two types of arson increase: the fraudulent fire for insurance profit, and the revenge arson, the "boyfriend's girlfriend" type of assault and murder with fire as the weapon.

Arson is by definition the non-accidental, deliberately set fire, for whatever reason: profit, passion, or madness.

In the profit arson, for the insurance, it can and does happen many ways. A merchant can be caught with his pants down in a fashion shift; he's got an overstock of peg pants, or straight legs, or flares—whatever—and the market shifts, and he's killed with his overstock. So he sets the fire to liquidate his capital. Arson seems to rise in inflationary times.

There's the respected citizen who has paid insurance for twenty-five years. He wants to retire but he can't sell his business or sell off twenty-five years of accumulated odds and ends—his merchandise, his stock. He convinces himself that after paying so many years of insurance without making a claim the insurance company owes him something back. He acts out his "right" to set fire to his stock—or house—and collect that insurance. Much like the car owner who had a fender-bender accident. He did it, he was driving, it was his fault, but he convinces himself the insurance company "owes" him and turns in a vandalism claim.

One landlord set fire to his own building after having numerous housing-court notices served on the property. He increased the insurance on the property, went inside in broad daylight with a can of gasoline and an armful of scrap lumber. He simply carried in wooden boxes, boards, kindling, spread it all around, set it on fire, and gutted the whole place. A tenant saw him do it, reported it, and we got a conviction before the man could collect.

Another man had a vacant building and the taxes went up on it one time too many. He had no insurance on it, so he called his insurance company, told them he wanted insurance. His agent told him the property was uninsurable. The man citing a lot of business with the agent, insisted on having a policy written on it and sent in on the possibility that the insurance company might approve it. So the agent wrote a $10,000 policy, attached a premium check for $66 to it, and sent it in. Before the policy came back *unapproved* six days later, a fire had gutted the place and the property owner collected. Since the policy had not been officially unapproved in the meantime, and since he had a binder with the company, he would have collected $10,000 worth of insurance if we hadn't proved arson. Not too smart of the company or agent.

Then there is the other type of arson: the revenge fire, the grudge fire, assault-with-a-deadly-weapon fire.

This guy got mad at his girlfriend because she was seeing another man. He decided to burn all her clothes, so he piled them up in front of her closet on the floor and set fire to them. He didn't really intend to burn the whole building down. He had no conception that a little fire can grow into a big fire. He only wanted to burn up her clothes.

There was a woman in a third-floor apartment, really angry because of the noise from the kids on the floor below. Drunk. To get at them, she threw a jar of kerosene down the stairwell and lit it. Can you believe that. *Down* the stairwell!

Males seem more likely to be involved in the big, multialarm arson cases, but women can be just as diabolical at setting fires as men. And children do their share: the playing-with-matches graduates, the sadists, the kids who start out with a prank, or an experiment, and it gets out of hand because they don't realize how far a little fire can go, or how fast.

There were some kids breaking into a moving and storage warehouse, through the roof, to get some bicycles. Somebody had told them there were bicycles in there. They took the skylight off and looked inside, but it was too dark to see. They were trying to decide whether to jump in blind when one of them lit some paper and tossed it in so they could see where to jump. As it happened, the paper landed in a barrelful of loose packing, a barrel filled with shredded paper and excelsior. We had a nine-alarm fire.

Now here's a real mind-bender. Down around the Baltimore Inner Harbor we had a lot of old, big, vacant warehouses, all of them condemned, just sitting there waiting to be razed for redevelopment. They were all boarded up.

Once a week for a long time, on the weekends, usually on Saturday nights, one of them would go up in flames, regular as clockwork. Three weeks in a row, then skip a week and so on. Over fifteen major fires in twenty weeks.

We searched every possibility and came up with nothing but theories and rejected most of those. One theory was that it was the wrecking contractor. What burns in fire? Wood. With dumps and landfill areas getting scarcer and more expensive by the week, a fire only leaves stone and concrete and steel and these are easy to dispose of compared to wood. But every wrecking contractor for miles around had a piece of that pie and there was no collusion, no plot among them.

It wasn't the owners; it wasn't for the insurance: most of these buildings were owned by the city. All of them were condemned, slated for demolition in the near future. No one was ever killed, but three firemen were seriously injured.

We were really scratching our heads and thinking it's got to be a professional. This is no amateur and it's more than a prank. Because of the pattern, the method, there's got to be a plan, a design, a purpose to it that we just can't figure.

Baltimore is a port. All of these were on the waterfront. Sailors, harbor pilots, someone connected with regular shipping could be involved. There was no attempt to vary the schedule. It had to be someone awfully good at it and someone awfully sure of himself. Maybe someone in on a boat for three weeks, then off to Norfolk or Philadelphia or New York for a week by water, then back and bam, bam, bam.

One of our captains was driving past one of these late on Saturday night. He thought he saw a flash, a small light up on the seventh floor. He stopped and backed up. Before he could get out—he said he had never seen anything go up so fast—that flash traveled from the seventh, up and down so fast, that the whole building was ablaze before he could call it in. He called four alarms right off the bat—

that's thirty trucks and 120 men, and they ended up calling five more alarms before it was over.

Personally, I prefer this theory; although it seems farfetched and was never proved, it answers all the questions. I think it was someone practicing a certain type of arson, training or being trained. I think somebody was teaching a class, or practicing for perfection.

And one graduation day, the lesson was over. He, or they, stopped as suddenly as he started and never came back. Maybe we'll never know.

In a paid fire department, arson investigation is much more sophisticated than in a volunteer fire department. The volunteer is a citizen, a businessman, a farmer who has a business to run. As a rule, he hasn't the time or the training to investigate the fires he fights. Often there isn't money for an arson investigation unit. So the volunteers put the fire out and then they go back to work. Heaven only knows what the real arson statistics are in some areas.

Insurance companies investigate to some degree as do police if there is any real evidence of foul play, but only the larger cities can really follow through. Even the large cities often don't have the funds or the manpower to do a proper job.

Arson is bad business. All the more reason to beef up your self-reliance and approach your fire from your own individual responsibility to escape and survive.

With all your personal respect, precaution, and prevention of your fire, it is possible that someone else could set it purposefully. It still boils down to your chances of escape, regardless of how it started. We'll catch the arsonist if we can and put him away if we have enough evidence. But in the meantime, you just get out—safely.

CHAPTER 23

To Fight or Flee

We tell everybody in the fire escape plan not to fight the fire. We tell you to use the fire extinguisher only to fight your way out, and then only if you have no other choice. There can be no other generality that can be safely made.

But there are some loopholes in that advice, some special exceptions where it would be reasonable for an individual to attempt some limited fire fighting, according to his own best judgment and his knowledge of his own ability and capacity.

It's a big country. Regions vary considerably. Local and specific conditions include everything you can think of and more. I am trying to help you to save your life, with your help. I am trying to show you the best way to do it. I do not want anyone to die because of his interpretation of this book. So I am forced to say, Don't fight your fire; don't even consider fighting your fire once you are out. My only purpose is to get you and your family out safely. To do this I have to say the hell with your house.

But I am not saying that you should just sit back and watch it burn. I am saying the fire department will put your fire out; they know how to do it and you don't; they're equipped and you're not.

Notify them the best way you can. If you have no phone or can't get to it, send someone to report from the nearest possible telephone or firebox. But there are many areas where the fire department may be twenty minutes getting there. This will probably be a volunteer unit. If you are in such an area and you completely understand that life is primary—including your own—it would be an assault on your common sense to tell you to stand around and watch the fire burn and spread.

Rural people are, as a rule, more self-reliant and more experienced in surviving at this level than city folks. Chances are if you are served by a volunteer fire department, some friend, relative, perhaps a member of your immediate family—perhaps you yourself—is or was a volunteer fire fighter.

At any rate, according to your abilities and experience, there are some things you can do to limit or extinguish your fire so long as you know how to minimize your own life-risk factor. Consider the role your wife and kids will play in this operation. They may be side by side with you, throwing that garden hose, that bucket of water, or that fire extinguisher at your fire.

Preplan this possibility. Absolutely. It has to be a common-sense thing. Don't race helter-skelter to put the fire out. If you haven't preplanned, if you're in shock or panic, no matter how desperate you are or how important it seems, leave it alone or you are back to exactly where you were before you saved yourselves. Your life is more important. If you are in an extreme situation of isolation or hazard, you'd better get together with your firemen right now—before your fire—to get all the training and advice you can get. This must be preplanned with the firemen, your family, and your neighbors if they are involved.

Never go back inside. Never fight an interior fire from the inside. Victims who go back in for anything don't always come out a second time uninjured. Fortunately, this does not apply literally to firemen; unfortunately, though, fire fighting is the single most hazardous occupation in the nation, perhaps the worst in terms of fatalities and injuries and this is where most firemen get it—inside that burning, smoking, collapsing building.

Look at it this way: You take too big a risk. It could be your fingers, your ears, your sight, a lung. Your family doesn't want that; you don't want that for yourself or for them. Think about it. You have no fire-protective clothing or air mask; you have no oxygen supply.

What can you do? Maybe you can keep it from spreading to other buildings; maybe you can contain it; maybe you can put it out from the outside. Maybe you can contain it until the fire fighters and apparatus arrive. Maybe you can do nothing.

Water puts out fires better than anything else (except for electrical or fuel fires). From a distance, apply water, from a hose if you have a hose and water pressure; apply it from buckets, anyway you can, from a safe distance.

Stay out of the smoke. Don't overexert yourself. Do not use a ladder except to attempt the rescue of human life, no matter how tempted you are.

Your clothing can burn you to death if it catches on fire. Wet yourself down if the temperature is not sub-zero. Wear a raincoat or storm gear. Shield yourself and your clothing.

Remember that things explode and that gases are poisonous. You know better than anyone what can blow up in your house. Think hard about that; make yourself remember. Remember you could have a heart attack or go into shock. Keep your distance from the fire.

Whether or not you can apply water, you can clear a fire break with a rake, shovel, tractor (but watch the gasoline tank); wet the tractor down, make sure you don't have a fuel leak, and keep your distance. You can remove flammables from the fire's path: real, probably, or possible. Wet down the fire break. Wet down adjacent buildings and property.

Beware the effect of wind and wind changes.

Beware of electrical wires and propane tanks.

Always have a clear exit no matter what explodes, no matter what the wind or fire decides to do.

Watch out for windows, doors, and even wall sections blowing out on you from the heat of the fire.

Make sure nothing falls or collapses on you. The only way you can be sure of that is to not get under anything, even for a second, that could fall or collapse.

CHAPTER 24

Using the "Right" Fire Extinguisher

Fire extinguishment involves removing fuel from the fire or making it unavailable for combustion; it involves cutting off the oxygen supply with foam, with steam from the water jet, or, by putting a cover or lid on the kitchen fire; it involves cooling the fire below its reignition temperature. And whether we really understand its principles or not, the application, by an amateur, of water to any small fire, except flammable liquids, gases, chemical or electrical, really works better than anything else.

Water cools the fire and the fuel, thereby "removing" the fuel from combustion. The steam from the water helps keep oxygen from it.

In saving your life, I did not mention the use of a fire extinguisher for any purpose other than fighting your way through or out of a fire as a last resort. Of course, I include fighting a clothing fire in this.

The purpose of the prefire planning is simply to get all potential human victims clear of the fire. Since fire accelerates so quickly, there is no time to do anything else. The smoke or heat detector gives you time to wake up, overcome panic and surprise, get your bearings, and get out.

But if you are already alert and awake, if you are right there when the fire starts—in the frying pan or the wastebasket—you have an opportunity to eliminate that fire then and there if you follow three simple rules:

1. Notify the fire department and alert any other persons to evacuate (*before you fight the fire*).
2. Don't attempt to extinguish anything but a *small* fire. If you can put it out within thirty seconds or so, attempt it, but start

yelling your head off to alert the others at the same time. Do not let your clothing catch on fire. If you find you can't do it *or* the fire increases—get out. Your exposure to smoke at this point will become seriously dangerous.

3. Whether you put it out or not, notify the fire department *immediately*. I am willing to say—in the same sense that there is no such thing as an unloaded gun—that there is no such thing as amateur fire extinguishment. Fire and fire extinguishment are profoundly complex. Regardless of your excitement, apprehension, or embarrassment, you need the services of the professional fire fighter.

Aside from smoky fires and flaming fires, there are basically three types of fire and three types of fire extinguishers.

The first and most frequent is the class-A fire, whose fuel is wood, paper, trash, clothing, mattresses, but *not* chemicals, petroleum products, or electrical fires.

The class-A fire extinguisher is water from a hose, or glass, or bucket, or from a portable extinguisher with a propellant or a hand pump to shoot it out the hose. This type of fire extinguisher may have helpful chemical fire-retardant additives in the water.

The class-B fire is fueled by flammable liquids—chemicals and petroleum products—in addition to the above combustible fuels. It includes kitchen grease, cooking oil, and gasoline. *It should never be fought with water*. Water will only spatter and spread it and burn you or cause an explosive chemical reaction. The best extinguisher for this is the type using inert gas or foam. CO_2 or a smiliar inhibitor is the chemical employed. This basically "puts a lid" on the fire to keep out oxygen. The third type of extinguisher—the DC or dry chemical, sometimes called the ABC type—does the same thing and is effective on this type of fire, but is not as effective as water on class-A fires.

The electrical fire is class C and *it also should never be fought with water*. The electrical fire can also include any or all of the combustible types previously mentioned. The first step in putting the electrical fire out is to disconnect the electricity: Pull the plug, unscrew the fuse, or trip the circuit breaker at the panel. The foam or gas extinguisher and the DC—dry chemical—extinguisher can put this kind of fire out only if the electricity is turned off. If disconnection is impossible, it can hold down the spread, but that is a job for firemen. In such a situation, you should be getting the hell out of there by now, closing all doors behind you, and not closing anyone else in.

The danger of fighting an electrical fire with water is the possibility of electrocution. Water—puddles, streams, mist, or steam—can transmit electricity to you.

If you have a fire extinguisher in the house, be sure it is located where everybody knows how to get at it—fast. Every new extinguisher has an instruction label. Read, understand, and memorize it into your preplan and be sure everyone else in your household does the same. A fire extinguisher that you can't operate is useless. Be sure now you understand how to trigger it; sometimes safety pins in the trigger may jam if you are unfamiliar with their operation.

You have water in the house. If you fight your life with that—and there are no petroleum products or electricity involved—this is all you need to know: Alert the others and notify the fire department. Stay low, out of the worst heat and smoke. Take short, cautious breaths. Keep an instant exit available; don't get cut off or trapped. If this is even remotely possible, you have no business being there. Get out and close all doors behind you. Don't close anybody else in. Beware explosions or flare-ups. Direct the stream of water at the base of the flame, not just aimlessly at the smoke. Remember fire and water make steam and steam can hurt you. Be as careful of this as you should be about all gases and smokes. Use a side-to-side sweeping motion of the water stream to cover and cool all the burning surface.

Sweep from the fire edge nearest you and progress toward the fire farthest away from you. With floor fires sweep from the outside edges inward toward the center of the fire. Sweep floor surfaces first and then progress upward to walls and vertical surfaces. Wet down any flammables the fire can spread to. If the ceiling is on fire, get out and close all doors. Stay on the outside of closets, pantries, small rooms, and attics and shoot the stream in. When you have knocked the fire down and nearly out, move in if the smoke and heat will let you. Pull apart the burned area with something other than your bare hands to get at the hot spots with the water.

Always check deep inside upholstery and bedding and apply water to all smoldering or hot spots.

Ventilate only after the fire is out; don't feed it oxygen.

Notify the fire department—*again.*

There should be a law requiring this. It is criminal not to have them make sure that any fire *is all the way out.*

All fire extinguishers (other than those using water) should have laboratory certified labels. They are not effective for putting out upholstery or mattress fires and, in any case, they should be used only to contain the fire until fire fighters arrive.

The best recommended, least expensive, most effective fire extinguisher with the broadest application, other than water, is the ABC —dry chemical—type. These and the CO_2 types start at the "beer can" size and are readily available and affordable in the 2¾ pound and 5 pound sizes; these are the weights of the contents and the containers add to that weight. But they are light enough for even a child to use.

Throwing bicarbonate of soda (baking soda) on the fire applies the same principle, if not the same chemical, to the fire. The DC type blows a white powder or dust similar to bicarbonate of soda on the fire; but remember *not* to use flour, baking powder, salt, talcum powder, or any other powder or dust on the fire or you may have a dust explosion.

I also said you can put a lid on the pot or pan or close the oven. Turn off the electricity or gas. Remove combustibles from the area, but never attempt to pick up or move that flaming pan of grease or burning food.

If you do attempt to move any hot fuel—such as smoking logs, a smoldering wastebasket, mattress, or clothes or lint from the dryer—out of the house, be very careful and get it *all the way out* onto paving or dirt. Do not put it in a hallway or on a porch or anything else that is flammable.

If your clothing is on fire, deal with it the same way as you would deal with any victim's clothing fire:

Don't run.
Get on the ground or floor.
Smother the flames to prevent deeper burning.
  Roll on the ground or roll in a blanket, rug, coat, or anything that is not more flammable than the clothing—obviously not a gauze curtain. Be sure to keep the victim's head outside the covering.
Remove clothing. If it sticks, leave it alone.

Wrap any burned part of the body in a clean, dry cloth—a sheet or towel—the cleaner the better.

Get a doctor fast. Call an ambulance!

The victim should remain lying down, calm, and warm.

If burns are serious, treat for shock until the doctor comes, as follows:

Keep the victim lying down.

Elevate the legs and hips and loosen the clothing.

Keep the victim warm, but do not overheat. Wrap blankets completely around his body, over and under.

Get that doctor or ambulance there—*quick!*

CHAPTER 25

Getting Involved

If you are a pedestrian or you are driving and you see a fire, what do you do? It's not your house or your fire but you must do something. What you do—quickly and correctly—may save lives. Or what you do may cost lives—even your own—depending on your judgment and your decisions and the speed with which you make those decisions.

Remember, if you go into that burning building, you are entering the statistical-probability situation of being injured or even killed. You have several advantages: You are fresh; you have not been breathing smoke; you have not been scared half to death.

Here is your fire preplanning for someone else's fire. Like many of the others reported before, this is a tough situation; like the others, it is your judgment that will direct your response.

Here's mine.

Don't do nothing. Do something! Do it at the level of sane commitment you feel you can afford.

(I'm thinking: How do you turn someone off on something this important? That person inside will die if you discourage his rescuer. Someone is in the building who could be saved if the right direction is given. Appropriate rules. I can't stand in the middle. I can't simply say *Do it*. I can't simply say *Don't do it*. This man could stand outside and do *nothing*. What should I tell him? I know the perils. I am outside with him. I am inside, dying, because he doesn't know for sure what to do.)

Either alert the occupants *or* notify the fire department the best and quickest way possible. If there are other "outsiders" involved, share these duties among yourselves. Don't delay if you are going to get involved; take charge if someone else doesn't. You can't order a passerby into a burning building but you can send him to summon help!

Suppose it is a smoldering building, a residence.

Nobody is at the windows and you can see no sign that anyone inside has discovered the fire or is trying to escape. There is no sign of any activity but that of the fire within the house.

Make a lot of noise. Pull right up to the house and lean on your horn. Yell, holler, scream—"FIRE!"

Do anything you can to make the most noise you can.

Don't throw anything through a window.

The front door may blow out on you if you open it or force it.

If you can wake a neighbor instantly, do it. Bang on his front door; break his door glass if you have to. Yell "*Fire*" into his house. Clearly identify yourself as wanting to report a fire. You don't want to be mistaken for an intruder. Don't get yourself shot.

If you can connect with a phone in his house, use it. You're awake, remember; he's asleep and you just yelled "*Fire*" into his house. He's confused and dazed. You use his phone if you can or make sure he does.

If there is no phone, calculate the farthest you might have to go to make the nearest connection with the fire department. Any telephone, any alarm box, any police car, any radio-controlled taxicab, any motorist who can report it.

Now obviously if the nearest phone or neighbor or fire department is miles away, you're going to have to decide how far you are willing to go. And I don't mean to report the fire. This calls for drastic measures. If you can't report it and you can't share the responsibility, it's between you and the victims. Try like hell to arouse them!

Now let me tell you what people usually do and what usually happens. A passerby, afoot or driving, sees a fire, stops, looks, and plunges right on in. He's going to be a hero, because he's seen it on television and in the movies so many times he knows it by heart. He doesn't take the time to evaluate the situation and *to evaluate himself*. The average guy is neither Clint Eastwood nor a fireman; he stands little chance of becoming a hero in a situation like this. If he enters a fully involved house, the odds are on his death or serious injury, *without saving anyone*. That goes for police, too. (The first duty of police, on the other hand, is to report the fire on their radio —no exceptions.)

So, the first thing this guy does is run to the door and try to smash it in . . . with his feet, his fists, a trash can, everything but his head. If he can't get it open or if he does and doesn't get blown across the street or burned to a crisp, the next thing he will do is head for all the windows and start breaking them out until one of them gets him.

He is trying to alert the persons inside; he is trying to rescue them and he is trying to "ventilate the house" because he saw it on "Emergency" or something, the difference being they wouldn't ventilate a burning house from the bottom first unless they were trying to kill the occupants on the upper floors, which they aren't.

This action intensifies the fire and further endangers the victims; the fire department will prefer to do its fire fighting on an intact house.

Then if he can find a way in, in he runs. At this point, he has about the same chance of surviving as someone inside the house, because by now he's eaten a lot of smoke and has probably been burned or bruised by what he's done so far.

I cannot sneer at this and I will not sneer at this. This man is to be commended for his guts and his concern; but he is approaching the situation like a fanatic, which can be defined as a person who "redoubled the effort and forgot the aim."

Now don't mistake me here. He should make every possible attempt to alert the occupants *from the outside* before he even considers going in or breaking holes in the house. It may very well be you will decide to do that at the end of your processes; but this guy did it at the very beginning of his. You must exhaust all your exterior possibilities first, before you even consider entry. Then, if you do decide to enter, do it rationally, sanely, and with full knowledge of what you might be getting into and of your capabilities to do it and succeed. Accept the consequences whatever they may be. The aim, which you must remember well, is that the only possible reason to enter is to save life, not to sacrifice your own needlessly, or otherwise. You have no air mask, your clothes are highly flammable, and I assume you have never done this before.

The only good fireman is the man who is afraid of fire. If he's not, he's crazy and he's got to be very young and inexperienced, because he won't live long with that attitude. The same goes for you.

Now, when should you consider going in? When you are absolutely convinced that someone is inside. When you are absolutely certain that there is no way left to alert them from the exterior.

If it is an impossible situation, you must not go in *no matter who or how many are in there*. That's not even brave; that's not even human; that's purely suicidal, and that's selfish. You could kill a fireman a half minute behind you, trying to save you. Because even if it's obviously impossible for you, he has a slight edge with his equipment and experience. But you may kill him, too, if he appraises the situation as not quite impossible.

There is not a good fireman alive who would walk into an impossible situation, who would sacrifice himself for a life he knows for a fact he cannot save. And there isn't a fire officer alive who would suggest, much less command, it.

But this is one of the reasons why fire fighters die at their job. They're dealing often with heavy odds and intuitive decisions, with a wealth of experience behind them, and sometimes they misjudge.

Size up the situation from where you are. There's going to be a lot of smoke in there. Preplan it now. What's the object? Obviously to save someone you know is in there. How bad is the smoke? Take another look. Where is the fire itself? How big is it? Where's it coming from? Walk around the building. Stand back and *see*; find out as much as you can.

Right now, firemen could arrive and you would have something valuable to tell them.

If they don't, it's going to be even more valuable to you.

OK, this is it. You are going to make a sensible entry. You know all you can find out from the exterior, and you are going to try to alert or rescue someone; you have accepted the responsibility.

Going in is like coming out. Feel the doors; test them going in as you would in your home coming out. The difference is that in the latter you are saving your own life because you must; in the former you're risking your life because you *feel* you must.

If the door responds properly, open it. Keep a fast exit open and remember to memorize your route in because that is your only known life line out. See that nothing on your way in will trap you on your way out.

Stay low. Don't overexert yourself.

Every step of the way requires the same up-to-date, realistic evaluation of what you find and how you interpret it. As you penetrate the building, your mind may start saying no; you may decide you can't make it any farther for reasons of endurance or hazard. When that happens, *get out!* No one could possibly ask more of you than what you have already done, even yourself, even if you happen to be a policeman. Have the common sense, the humanity, to get out. Escape when you think you cannot go farther. Limit the death to one instead of two, six instead of seven. Escape while you can. You have valuable information to give the fire department, information that you didn't know before, information that may save the life you could not possibly reach.

There are so many variables in each fire situation. In the city, we respond within one minute to two minutes; in rural areas, help may be twenty minutes or half an hour away. You are a neighbor, and you can't let these people burn. But what right have you to burn with them!

Only you know the building you have just entered; you know this is not a movie; you are not in a theater. I cannot tell you how far to go, but I do tell you, don't go farther than your confidence! You are there, and you decide. If there is any doubt at any point, *get out!*

CHAPTER 26

Alert at Work or Play

Fire may catch one by surprise and certainly one can be overcome instantly by an explosion at anytime.

Awake and alert you are physically and mentally prepared to withstand, extinguish, or escape your fire. In this state, you have the same chance of mastering your fate in an encounter with fire as you would in preventing and surviving a fall, an auto accident, or a wartime emergency. Some of it is up to you, and some of it isn't.

The first rule is, don't do anything stupid. The second rule is, discover and assume your deepest level of personal responsibility. Preplan defensive living, the same way you would approach defensive driving. The more alcohol, drugs, and drowsiness affect your alertness, the more likely you are to become a victim.

The goal is not to keep yourself in a perpetual state of self-induced paranoia. One can only go to certain lengths to survive physically; beyond that point, it then becomes paramount to withstand and survive the paranoia, because that in itself would be the greatest threat to life and life enhancement.

When you are alert, survival response to fire involves a heightened sense of common sense, an enlightened, practical responsibility and alertness. It assumes a heightened awareness of probabilities and a general, rather than fanatical, preparation of the skills and means to survive. You cannot afford to assume or predict the co-operation of anyone else to effect your survival. It requires attention and practice.

You must approach each problem as if each other participant will behave foolishly. Your most effective attitude is to act in such a way that your life and survival are enhanced by your own actions regardless of any external help. Needless to say, if the other guy responds rationally, be quick to accept his help and profit promptly from its good effect as a fact. But no matter what anyone else contributes to the problem—good or bad for you—you must constantly approach the problem as your own unsharable responsibility.

Part of your problem involves the survival of the other participants; in making sure they don't deter you, you must at the same time not be a detriment to them. Your error could be fatal to yourself and to the other, just as their error could be fatal to you.

Assume difficulty; accept ease.

The sleeping fire situation involves your personal responsibility to survive fire as an individual and as a family, basically in your home, but also in any sleeping situation in a temporary "home" such as a motel, a friend's house, a tent, a camper, or a Pullman car.

Here too, this involves preplanning your best response to fire at any time in any place. Since you will never experience all exterior situations it is not necessary to get too exotically far afield in the imagination to establish your core complex of probable situations.

Awake, alert, you are familiar with your home and your preplan; likewise, you know your car and your place of work and there are certain common factors to being alert in any movie, restaurant, or church.

You must constantly size up your surroundings in terms of immediate and potential hazard. In familiar surroundings, this is relatively easy to preplan but the possibility of fire in unfamiliar surroundings must be generally preplanned as well. The principles are generally the same, common to all situations.

You must use your imagination to enhance the predictability of personal hazard, and you must use your common sense to enhance your survival. The best fire plan is prevention, but you must assume fire to be highly probable. No matter what you may have done to prevent fire, it may be caused by another person, by an "accident," or even by your inadvertence.

I am trying to establish the fact that there is a high survival attitude that does not derive from paranoia nor induce it; there is an attitude that gives you the greatest possible chance of survival. Some people are truly accident-prone; statistically, this is possible even without their subconscious willingness to set themselves up. In actual practice, the individual can change his odds by preplanning his response to the high incidence of accidents or coincidences. If he keeps falling from ladders, no matter what precautions he takes, he should assume that nature is trying to tell him something; whether he comprehends the message or not, he should be able to assume that if he doesn't climb a ladder, any ladder, he won't fall off it.

Avoid high hazard situations.

Be alert to potential hazard.

You are better off assuming that your fire will indeed happen. By even a casual set of planned responses, you can reduce your statistical probability of becoming a fire victim by at least 50 per cent. You can reduce your odds even more by continuing to take the possibility seriously. Your preplanning will then be as real to you as any fire because you vividly assume the reality of the fire which has not yet happened. Then you are not going through the motions of a fire drill; you are, in reality, escaping a real fire each time you practice your preplanning. Each rehearsal of a play is real; the play is real; the actors are real. Without these things taken as reality, opening night will be a bomb. If you assume opening night in your rehearsal for it, it is highly likely it will go well if it ever comes. If, in the back of your mind, you do not believe opening day will ever come, God help you when it really does.

We've been dealing with the fire in your home and your relationship to it. Obviously, if you have a good prefire plan and a good evacuation plan going for your home, it will be easy to carry these principles over into your work situation.

Most people work in offices, factories, stores, or warehouses. Students do their work in schools.

Office fires are caused by many factors; the more people in any situation, the more varied the possibilities and the higher the chances of fire incidents.

These fires start in so many different ways: in wastebaskets, trash cans, storage areas, around electric coffee pots (even at night when they are plugged in), ashtrays, copying machines, electrical office equipment, lighting fixtures, and air conditioners. Some of these things make a hell of an odor; all of them make a lot of smoke.

There are also many ways the fire and smoke can travel: hallways, heating and air-conditioning ducts, utility and elevator shafts, concealed vertical and horizontal voids or dead-air spaces, stairwells, and so on.

Generally the responsibility to unplug the coffee pot and to smoke carefully is the employee's own responsibility. Generally the condition of the building itself is the responsibility of the landlord or employer. His responsibilities extend to providing safe exits, basic fire extinguishment equipment, and training in its use, as well as basic prefire planning and fire evacuation planning complete with a building fire warden and floor fire wardens who co-ordinate all planning among employers, employees, and the fire department.

It is the employee's responsibility to enter into this fire planning as fully and actively as possible and then respond and obey this planning to the letter in the event of fire.

Even granting the possibility of a building approaching the ideal fireproof condition, I'm reminded of the comment of a fire fighter after a nursing-home fire, "It was as safe as a furnace, but everything in it burned."

No matter how sound your office building, most of your furniture, carpeting, wall covering, floor and ceiling tiles, veneers, waste and scrap, computer tape, microfilm, supplies and materials, equipment and machines are not "fireproof." Many of the plastics, vinyls, synthetics, foams, electrical insulation, and so on, not only burn, but give off bad poisonous gases, which can be more lethal than the fire itself.

Once these toxic gases start moving, they can enter or get pulled into so many different air-movement systems so quickly that they can reappear in lethal concentration anywhere at all in the building, near or distant from the fire itself. And they are also going to tend to move upward and permeate everything as they rise.

Here, because you have the possibility of so many people concentrated in any large office building, the ingredients for panic are many.

The legal requirement and enforcement of prefire planning and fire evacuation planning in offices and other work situations as well as schools and the like is rightfully on the increase.

The basic responsibility unit remains the individual, as always, but the presence of the building fire warden and his subordinates establishes a very healthy and effective chain of command. As in your home fire planning, all of this must be seriously considered and actively practiced.

The fire warden should be trained by the fire department at the employer's expense. A fire warden should be present in the building whenever occupants are present. If he checks out of the building, his next in command must check in and remain present until he in turn is replaced. There should always be someone present who is ready, willing, and able to take charge of the possible fire situation.

The duties of each floor or subwarden include regular inspection of all exit routes and doorways and all exit signs and lights.

He is in charge of organizing the floor or unit planning and he is the local fire "boss" in the event of fire and fire evacuation. It is his responsibility to make sure everyone is out of his area. That includes assigning buddy relationships to secure the safe, speedy escape of all visitors and handicapped personnel in his area; he is the last to leave his area and is additionally responsible for accounting for all persons at the assembly point and for reporting to the fire department whether all persons in his charge are accounted for.

He is in charge of sounding the alarm and assuring full alert of the fire to all persons in his area. Incidentally, some of the newer installations broadcast a prerecorded calm voice to announce the fire alarm so as to minimize the panic reaction; then a strong voice gives the directions. Those directions spell out the simplified elements of the preplanned fire evacuation procedure—the exact steps to take at that time. Whatever the building and floor-level alarm system, it is the floor warden's responsibility to see that the message is received and understood by everyone there. If any alarm system should malfunction, he must transmit or delegate transmission of the alarm to all persons. This must be loud and thorough enough to alert everyone there, regardless of whether it is an open floor or closed-cubicle situation.

In any case the persons present have to be psychologically prepared, instructed, and trained in the preplanning in how to proceed from that alert or alarm onward without hesitation to their full escape—to outside, the roof, or to whatever predetermined destination has been assigned them by the fire wardens.

Erratic or individualistic behavior in mass-fire evacuation can kill a lot of people. The fire planning and evacuation objectives are exactly the same as in your home plan except that the basic unit here is more social than personal. You *must* completely co-ordinate your escape with that of others or the chaos, panic, and bedlam that could result from erratic individualism can kill as many people as fire and smoke.

The time to air any divergent views is in the preplanning meetings and practices, but be sure your additions to the discussion contribute to the planning. Accept the plan finally evolved. Of course, nobody but you can force you to work there and submit to the best mass fire-exit plan available for that situation. Good old Harry Truman: "If you can't stand the heat, stay out of the kitchen."

Building wardens should report to the fire department regularly and share appointed inspections with them. They should consider and act on all reasonable instructions fed back by the fire department.

The fire department is especially eager to offer full services at this level, because high-rise buildings cause firemen enormous problems; they are best solved at the prevention and preplanning level. Manpower requirements for fighting a high-rise fire are triple that of an ordinary fire. Equipment limitations are serious. Most ladders made cannot reach higher than the tenth floor of any building. Nor can water-tower jets reach much higher than the eighth floor and they drastically lose effectiveness as they near their limit.

Floor or unit wardens should meet regularly with the building warden, but not at such a time that will leave any floor or unit unprotected.

Participate fully in your fire plan; exchange questions, answers, and opinions with everyone involved, but accept the final plan and submit to it.

You had better think about this—now. This decision to accept, tolerate, and obey in this situation *is* the very heart of that responsibility we have been talking about throughout this report. This is your responsibility and the core of your most successful prefire and evacuation planning. In the family situation, the greatest survival for all family members lies in the most complete surrender to each by all the others of any claim or dependency in the moment of crisis. Each must free the other *and himself* to the greatest, fullest possibility of the survival status of the loner.

Here, it is the *and himself* factor I am addressing, that ego or fantasy or claustrophobic component that resists or resents or rebels against any ant- or bee- or sheeplike role that may be assigned to him. You may come into work like a lion, but, if there is a fire, it is your deepest responsibility to come out of it like a lamb.

Forget you ever saw *The Towering Inferno*. I mean *eradicate* it!

The floor warden is responsible for the security of his assigned area; that means he is the last one to leave and is responsible for securing the valuables left behind.

These valuables—cash, receivables, charge slips, irreplaceable records, and things of this sort—may be lying around all over the place when that alarm goes off. They may be on desks, tables, shelves, anywhere and everywhere. How much time will be devoted to securing them? All these valuables are in the same status as personal valuables in the home.

No employer should ever place any employee in the position where his escape would be impaired or jeopardized for the sake of saving records or other valuables.

Now in counterposition to that, consider realistically that the neatest fire fighter in the world who comes into that room to fight the fire will probably either destroy by water or move around those things to a point where they can no longer be usable.

All things considered, it should be preplanned that certain responsible individuals voluntarily gather and drop all these things into some kind of fireproof drawer or safe in the available time between the sounding of the alarm and the exit of the floor or unit warden—the very last to leave. This way the orderly, smooth flow of evacuees remains unbroken and uninterrupted. As a matter of efficient office procedure, no more records should be strewn about at any one time than can be effectively recovered and protected within this brief period. In no instance should the fire warden's exit be delayed by this activity. When he says that's it, git. He must be the last one out for many reasons, one of them being an accurate accounting for the exit of all present. Another reason is that he is the "law in the plan," the security of these things as well as people, until he transfers this responsibility to entering firemen at any level in the building.

This should also be preplanned and practiced until it works without a hitch. If the security thing doesn't work, it should be abandoned because life is involved.

Any security guard assigned to that floor or to any particular business or office on that floor should normally surrender his authority and responsibility to the floor warden at the door and exit before or with him. He is neither authorized nor expected to jeopardize his life any more than any other person. In any fire chain of command, he becomes subordinate to the floor warden at the door and must leave according to preplan at the floor warden's request.

The duty for protecting his assigned assets is literally and fully taken out of his hands by the floor warden.

Once the fire department takes over the building and that floor, it is over. The fire department then assumes the job of policing the structure. Sometimes a fireman is stationed on each floor to make sure no one enters until the fire warden returns.

We—the fire fighters—don't know the people and we don't want unauthorized persons in wrong apartments or offices for any reason. We take full command of any building in a fire situation, in conjunction with the police department, until the fire is out.

Once out of the building, at a preplanned assembly point with his assigned group, the floor warden accounts for all present and all missing and reports this to the fire department.

The building warden is in charge for all of this happening as successfully as possible. He and the landlord are accountable to the fire department for locating all maps, plans, drawings, and blueprints of the building and for pointing out the location of important building features. In the preplanning, this should be one of the constantly updated exchanges between himself and the fire department.

The assembly or head count is absolutely important for obvious reasons. If there is anyone in the floor warden's group with current symptoms of heart attack, shortness of breath, shock, a pain in the chest, epilepsy, emphysema, smoke inhalation, diabetes, burns, or anything of a serious nature that *may* require medical attention, it is the floor warden's responsibility to bring it to the attention of the fire department. It is the victim's responsibility to bring it to the floor warden's attention. Likewise, persons with a history of an incapacitating condition must be looked out for.

If that person is you, don't wait, don't be embarrassed—tell somebody about it. Ask. If it is quicker to tell a policeman or a fireman or your floor warden or an ambulance driver—do it immediately. A radio contact will promptly dispatch to you insulin, medication, medical treatment, emergency service, hospitalization, a baby bottle, whatever is required. Ask.

If someone's health or ability is restricted, a buddy should be assigned to assist in his escape if he needs help; not to carry him out physically, but to stick close to him and offer reasonable assistance.

There's a problem if someone collapses on the way out that the floor warden will have to solve. The options are to carry or help carry the distressed person or to leave him as comfortable as possible and immediately tell the first fireman where that person can be rescued. I can't answer this one. It's up to the fire warden and his common sense and best judgment.

If the fire warden isn't there and your assigned buddy collapses, then that judgment is up to you.

So far we have assumed that the alarm signaled a fire *not on your floor.* If the fire is on your floor, the first person detecting it should sound the alarm and alert the floor warden who then makes sure all on his floor are alerted. Further alerting of the fire should be continued all along the escape route in such a way as *not* to cause panic or to slow down the orderly escape movement.

Needless to say, all such buildings should have detectors and internal alarm systems that are connected directly to the fire department.

With what you now know, you're in a damned better position to survive and assist in survival than you ever were before.

Here it is:

1. The alarm goes off.
2. Alert others.
3. Proceed to escape immediately.
4. Test all doors. (Are windows a possibility?)
5. Do not use the elevator.
6. If there is smoke, stay low.
7. Do not overexert yourself.
8. Enter the stairwell.
9. Close the door after you.
10. Proceed up or down according to previous instruction. (If no instruction, proceed down if the way is clear.)
11. Walk; don't run.
12. Stay calm; assure others.
13. Exit to the outside (at ground level or roof level).
14. Assemble.
15. Notify the fire department. (If they are already there, account to them for all present and any missing.)
16. Don't disappear.
17. Assist if requested.

See, you didn't have time to panic.

At no point have I recommended fighting the fire or even investigating it. That's not your job unless it's *your* wastebasket right in front of you. It's not the floor warden's job to fight fire either unless it's his personal wastebasket or ashtray.

His only fire job is to prevent fire, to preplan fire evacuation, and to take charge of any actual fire evacuation.

In a fire of this type the only persons—other than the fire department itself—with any right or responsibility to fight that fire are those people who have been trained and instructed to fight fire in your building by the fire department.

As for the landlord, if he's given you a safe building, thank him; he's spent a lot of money on it. If it's a fire trap, bitch, bitch, bitch until he does something about it; if he doesn't respond, move out and let him know why.

The worst school fire on record caused 294 deaths. It was in New London, Texas, in 1937. Gas, escaping from piping installed by a janitor, exploded at an electric arc.

In 1908, an Ohio school fire claimed 175 lives, a result of panic when excited children jammed up in a vestibule.

In 1958 in Chicago, 95 died as a result of a rubbish fire that started in the basement and spread up an open stairway into the third-floor corridor, trapping students in their classrooms.

Other large multiple-death school fires of this century became large scale tragedies due to:

Inward-swinging exit doors.

The blowing up of a school by a mentally deranged man.

Nonpractice of fire drills and nonperformance of practiced fire drills.

Combustible Christmas decorations.

A candle.

Rapid flame spread through combustible ceiling area that was not fire-stopped.

Jamming at exits.

Only exit blocked by fire.

These actual occurrences do not begin to touch all of the possibilities. In order to be effective, whatever precautions and planning are necessary in a single residence must be multiplied by the number of students and teachers in a school situation.

A recent two-year period involved 155 school fires. Life was lost, injuries were suffered, and precious education was disrupted. A large percentage of school fires are of arson/incendiary origin. Almost a third are caused by faulty or misused electrical wiring and equipment. Other school fire causes involve spontaneous ignition, improper storage and use of flammables and ignition sources, and general stupidity and carelessness.

Ideally all schools should have smoke detectors and sprinkler systems wired into an internal alarm and the nearest fire department. All schools should have intrusion alarms (separate and capable of being overriden by fire alarm) and security guards functioning twenty-four hours a day, every day of the year. About 75 per cent of school fires occur between 3 P.M. and 6 A.M., when the schools are unoccupied, and the security precautions—barred windows, padlocked fire escapes, permanently closed windows—provoked by this fact dilute the plans to evacuate the building when fires occur during occupancy.

All schools should have well-organized evacuation plans and drills, enough accessible exits, and alarm and warden systems practiced and efficient enough to minimize panic and permit immediate, uninterrupted escape from the building as quickly as possible following the outbreak of fire.

These fire evacuation practices are required in schools but normally fall short of the motivation and imagination required to make them most effective and productive. Even at the letter-of-the-law level they are necessary to evacuate people from the buildings through proper preplanned routes and fire exits. As on the individual level, they should be physically practiced frequently enough to familiarize all occupants with the escape plan, procedures, routes, and exits. Conduct during drills should become a matter of accepted routine, thereby minimizing the chance of panic. Practices should be carried out at unexpected times and under conditions which effectively and safely simulate the real fire conditions. In some cases, for the purpose of practice drilling, smoke pots have been used and, in others, primary escape routes have been blocked forcing the use of alternate routes.

Unfortunately, open stairways, inadequate exits, blocked and obstacled escape routes, structural hazards and flaws, and the presence of flammables still exist. And the level and prevalence is such that it is a miracle that more people don't die. Any school fire death is one too many. If you have any reason to believe your child's school is not fire safe, do something about it immediately.

The fire escape and doors must be clear, accessible, and easily *opened* from the inside. I don't care what other kind of problem your school has; it has got a mass-multiple-fire-death possibility too high to comprehend if those doors are locked.

Office, factory, and school situations are very nearly the same. In

EXIT
J.S.

the school situation the principal, or his or her equal, is the building warden responsible over the subwardens. He is under the supervision of the fire department in any fire-related or fire-safety-related matters. In this, he is responsible and answerable to the fire department —and ultimately to you and your child.

The teacher is the subwarden, responsible under the building warden, for the children's safe escape from fire. He or she is also answerable to the parent, the principal, and to the fire department for the best execution of that responsibility.

Just about anything that can be wrong with any building can be wrong with a school. In some of our educational buildings can be found minimum intelligence, maximum carelessness, and fire traps combined with countless variations of the old fire trap.

It is not the job of the building warden, principal, subwarden, floor warden, or teacher to fight any fire in any way *except* to fight a path out from the fire and to extinguish the burning clothes of any person.

In another context, I have always heard that bad politicians are elected by good citizens who don't vote. But if you spend election day driving stranded, invalid, and crippled voters to the polls, you can produce hundreds of votes rather than just one. So it is with principals and teachers whose single, solitary job it is to alert and bring all lives entrusted in their care safely out of that fire to a preplanned assembly point and there account for all of their charges.

At that point the fire department should be notified by one of them, if it has not definitely for damned sure been notified by someone whose preplanned assigned job was to do that.

Anything more than that should have been accomplished in the prefire planning. The prefire plan and the fire evacuation plan must utilize what is actually available in the quickest time possible. You can do only what is possible; you can use only what is available when the fire strikes.

For now, the school situation is the same as any individual fire plan on a larger scale, with the larger scale compensated for by the warden/principal and the subwarden teachers moving their groups in the orderly manner that would be expected of the individual in a home fire. Truly individual escape and responsibility, as in the home situation, would be chaos and disaster here, especially so because of the younger age and relative immaturity of the students.

Individualism should come in at the preplanning level, where everyone should be encouraged to contribute thoughts, suggestions, and general views on the situation and should receive in exchange, full and accurate answers to their questions and resolutions to their problems. But all this should be done before the fire!

When that fire hits, obey without hesitation; the evacuation plan is the only way out.

Obey. It is absolutely your best chance.

Have you ever seen a chicken farm? Thousands of chickens in them. Have you ever seen those signs they put along the road or on the driveway leading into one of those places? The sign says that if you blow your horn, you will kill a lot of chickens. One little beep at the wrong time and you will become a mass chicken murderer, whether you want to be or not.

Chickens live in very crowded quarters. They are very easily spooked, very quick to panic. It's hard to say what does it, what triggers them, but if you haven't seen chickens stampede, you've never seen a stampede.

It's worse for chickens because they really have no place to go. Not even a cliff to stampede over like bison or lemmings. No way to dive down suddenly like a school of fish. No way up like a flock of ducks. Only four walls, and each other. When 5,000 chickens pile up in a corner big enough for fifty, that's probably 4,950 dead chickens and maybe fifty live ones.

There are also warnings that you shouldn't yell "fire" in a crowded theater. It's nice to go to a movie or theater. If you like it, it's a real pleasure. There are some things you shouldn't do when you go to a movie. And some things you definitely should do, if you want to avoid the old barbecued-chicken condition.

Signs in movies post the seating capacity. Obviously more people cannot be seated than there are seats. It is absolutely against the law for overflow people to stand in the aisles or elsewhere inside the theater. In the lobby, the line waiting to go in must occupy a certain area only, behind a thick rope between chrome-plated holders.

Other signs say NO SMOKING and often no one does. Ushers as a rule are not trained in any fire procedure and are, as a rule, not big enough to enforce the NO SMOKING signs even if they were paid enough to risk it. The real manager is often not present.

The exit signs required at every exit must be suitably lighted and legible.

You have probably noticed, when you are waiting to get out of a movie theater after the feature, that the crowd even under normal, calm conditions, does not permit you to move very rapidly. And you have probably noticed "a little tension is in the air."

Imagine: If there were a fire or someone just yelled "Fire"—honked the horn so to speak—and those same people were trying to make their desperate exit under those conditions. It is quite probable that you would have a pileup or jamup—a stampede of people at the usual corridor exits. In this situation, it would be tough even to get to the aisle unless your seat was on, or near, the aisle. If some other fool wants to sit in the center row of the theater, halfway down, surrounded by people who just aren't going to be able to get up and move, then that's his decision. Aisle seats are safer. (Or are they? Movie theater aisles and seat fabrics and carpeting are not very well designed for people safety.) But everybody can't sit aisleside on a Saturday night or Sunday; there's a lot more room most week nights, and there will be a lot more room after the sparkle has worn off the first-run flicks.

If you find yourself in a situation that you are not comfortable with, get up and get out. No movie's that good; if it really is, it will be just as good in a month or two when you can view it in complete comfort.

Don't stay in any public place you can't get out of if people start to panic.

You may blow three dollars—or six—but if you are not comfortable, if you are not satisfied with the situation, do something about it before you get trapped.

We had about two thousand people in one theater and maybe three thousand waiting outside when we got a report of a fire in it. Nobody knew whether to evacuate the place or not and the stupid "manager" decided he wasn't going to evacuate the place until the owner told him it was OK.

In most places where there is a fire or suspicion of fire, a fireman is in command of the situation as soon as he enters the building. That authority usually has the sanction of the law. As it turned out there was no real fire in this case, just a lot of smoking going on in the men's room. But that idiot of a manager could have killed a lot of people.

There should be training of all theater management and personnel in protection of the building itself, but most importantly in fire safety and crowd control. A good theater manager can be great; dur-

EXIT
J.S.

ing the 1968 riots we ordered a particular theater evacuated, and the manager calmly stated over the sound system that we considered it best to discontinue the movie at that point. The picture was still running but without the sound track. A soundless movie screen is a sobering experience, sort of Gut-level Basic.

The manager's voice continued to say that he was going to turn up the lights and would everyone please exit in an orderly fashion. There was no fire and no cause for alarm. They would be given a rain-check ticket by the ushers at the exits. When the lights came fully on, the image on the screen disappeared.

That theater was packed, but everybody did exactly what the manager told them to do. They rose very calmly and began to leave as if they were leaving church. He said there was no fire and no cause for alarm. So, while I watched this evacuation, I started to wonder: What if there had been a fire? I had a long time to dwell on that because there were a lot of people going out.

When the manager gave the rain check at the door, many asked him why he had made the announcement.

He said, "The riots are reported to be coming this way; there have been some incidents."

Then I thought: What if he had just blacked out the screen and blurted "The riots are coming. Get out"? He would have killed some people sure as hell! That man did a beautiful job of crowd control. I thanked him for that.

The same attitude is equally healthy in the legitimate theater. There are maybe a few different problems like live flames in a stage production, actors with cigarettes, flammable scenery and props, but all of this should have prior clearance and performance surveillance by the fire department.

Theaters should have sprinkler systems, especially backstage, and an automatic fire curtain that can drop quickly. A theater so equipped would have a little more leeway than an unprotected one.

As a patron, check out all those exits; sit on the aisle and if anything looks wrong about exit access, even due to crowds in the lobby, I strongly urge you to have a heart-to-heart talk with the manager.

Tents are going out of style now, but they used to cause big fires in their heyday. In the South and the West you still see them: gospel tents, revival tents, and once in a while a real circus tent. They're not supposed to be highly flammable, but many of them are. So are the wooden bleachers or chairs in them, and the sawdust on the ground, and all those Sunday-go-to-meeting clothes of the audience.

That's why in many places you have to get a permit before you can erect a tent that is over 120 square feet in area. This gives us a crack at regulating and inspecting how it is erected and the chance to make sure its material is properly treated for flame retardancy.

In the same vein, propane or bottled gas stoves should be inspected frequently by fire people to make sure they are in good, safe operating order.

There are lots of automobile shows, boat shows, and similar exhibitions where gasoline-powered vehicles, motors, and tools are displayed. Demonstrations of these vehicles are out of the question, as far as I'm concerned, if the exhibit is inside. All gasoline tanks must be drained somewhere away from the auditorium before they are brought inside. All gas tanks drained and purged with carbon dioxide, CO_2, before they are brought inside a public place anywhere. And the batteries disconnected or removed.

Crowd control is important also in these large indoor places, auditoriums, civic centers and the like. Firemen and policemen should be stationed there for the run of the event.

The large stores should notify the fire department when they expect superlarge crowds. Perhaps occupancy limits should be enforced by some kind of crowd control at the door, limiting the number of people inside at any given time. First come first served. Then let them out and let the next wave in.

Bingo halls have occasional overcrowding, especially by older people. They're less agile, less quick, but they smoke like youngsters. There should be more provision for supervision in these places with everything really clearly marked and spelled out. We had a bingo fire a couple of years ago. Two dead, as I remember. One of the ladies got all the way out and then returned for her fur coat. There are church suppers, bazaars, oyster roasts, bull roasts, and the like held in the basements of churches.

It really pays to get the feel of any place you enter, and get out if it feels as if it could get too hot. Complain if you feel you should; certainly if you feel it's critical.

Churches are wide open for arson, literally, and thieves and vandals. Loss is primarily of property; relatively few lives are lost because churches are unoccupied so much of the time and most incendiary fires are set at night.

What concerns us about life loss in church situations is the use of open flames in the church by the congregation, not just on the altar. Many churches allow their parishoners to use candles in open processions and services within the church building. This is very dangerous and a subsequent fire—including a clothing fire—could not be ascribed solely to the will of God.

A church or synagogue has to be as safely maintained as any other public building where large numbers of persons congregate.

Particularly here because so many aged and very young persons are involved in services and functions, especially in its social and educational events. Safety precautions must extend to the sanctuary, rectory, church school, and any related structures and gathering places, as well as the church itself.

Many churches are very old and very valuable from historical and architectural perspectives. There is a traditional resistance to changing any visible part of them, even to conform to the newer fire codes and laws. But a church holds many people, over and over again for many, many years. There can be no special exemptions of any kind for any sort of building that houses large groups of people at any time. If lightning rods can blend into the architecture, so can smoke detectors, sprinkler systems, and other protective devices.

Where open-flame ceremonies are still permitted, it would be good for all concerned to have the eye of the fire department on the ceremony. This means giving prior notice to the local fire people so they can be on standby if not on hand. Present in spirit, let us say, if not in the flesh.

In those services where altar candles are involved, they must be under the direct supervision of the priest conducting the ceremony. He is directly responsible for the highly flammable altar boys who light, carry, and snuff the candles.

He is also responsible for seeing that all candles are extinguished at the end of the service.

We had a procession with two hundred people parading the streets: men, women, children, old people, each with a live candle. Doting parents, mothers giving them to their children saying, "Here, hold this." I was nervous the whole time but we were lucky and, if they ever suggest it again, I am going to veto it automatically. Live flames are dangerous; it's that simple and that deadly.

But then some religions handle live rattlesnakes.

If you attend church or any services where you experience discomfort from candles or anything out of the ordinary, explain your discomfort to the clergymen.

At any rate, the escape plan from a church is the same as anywhere else.

Check the exits; preplan.
If there is fire, sound the alert and bail out.
Assemble outside.
Notify the fire department.

At a recent four-car collision, a number of people were injured; some of them were pinned inside and others were waiting to be removed by paramedics in the ambulance crews. It is important not to touch anybody injured in a car; you could break their spines getting them out.

Anyway, gasoline was leaking, flowing into the gutter and there were thirty, maybe forty, people standing in that gutter or next to it, along the curb. Some of them were smoking. I'm sure it didn't even cross their minds that we could have needed ten or fifteen more ambulances real quick if the gutter exploded. A fireman foamed the gutter and under the car and moved the people along. They were lucky.

Another one wasn't so lucky. A kid deliberately threw a match into the gutter—with gasoline overflowing in it—and the entire truck went up and people were injured. We made a civil arrest on that occasion.

If you're in an automobile accident, you should get out of the car as quickly as possible. And move away from it, especially if the car or any other car near it shows any evidence of being physically damaged in the motor or gas-tank area. The fuel system can rupture at any point, even from impact alone. Do not re-enter the car or use it as a seat or bench. Your car has *gasoline* in it, remember. Don't let anybody pull you out of it, if you are stuck or badly injured; but do your best to get out if the car is burning or there is gasoline all over the place. Turn the motor off. Don't smoke and don't let anybody else smoke anywhere near the car. Disconnect the battery, if possible. You should have a CO_2 or dry chemical (DC) extinguisher in the car; if one is available, use it on the fire if it is small enough to approach it *from the exterior*; if not, let it burn. If there is no fire, keep the extinguisher on standby until the fire department arrives.

The rules are basically the same here as in any fire situation. All escape from the car or truck. Assemble. Notify the fire department, or the police; the police will automatically notify the fire department and call for an ambulance.

Not all motor vehicle fines result from collisions or moving accidents. Smoking in "bed" is just as dangerous in a truck sleeping compartment or taking forty winks of shuteye on the seat in a car or truck. Car and truck upholstery is just as flammable as any bedding in the home.

One man I know of started a smoldering fire with a cigarette in his rear-seat cushions. He "put the fire out" with a little water—two days in a row.

On the third day he was at a picnic and locked the car up tight and left it parked. When he returned, the "extinguished" fire had totally burned out the inside of his car.

There's a lot of "exotic fuel" in any car or truck, in the paneling, upholstery, wiring, and accessories. The fumes can be deadly. In the event of gasoline or electrical fires here, as anywhere else, do not use water for extinguishment. Disconnect the power source—the battery, usually—and then use CO_2 or DC.

Beware gasoline running down a gutter and into a sewer or other confined space. This situation has the explosive capacity to put a telephone pole a block away into orbit. Beware and stay clear of the accidents of others, even the accidents "about to happen."

When a truck has a sign reading FLAMMABLE or INFLAMMABLE (same thing) or EXPLOSIVES on it, it means exactly what it says.

Most of these trucks are restricted as to routes and in most cities they are not even allowed to park; but they can have a breakdown anywhere and they travel the same highways you do and stop at the same restaurants, motels, and traffic lights. Give those things a wide berth; they are dynamite, sometimes literally that.

It is a fair rule of the road, whether you are in the city or out on the highway, to give that gasoline tanker and the like plenty of room. You may not be happy with his lane-changing all around you, but you don't pick a fight with an atomic bomb.

You can also put a good bit of confidence in that truck driver's likelihood of being fully aware of his potential danger. But keep a respectful distance from his cargo, moving or parked. If moving, and you see something going wrong with his truck or cargo, try to signal the driver, but don't tailgate or ram him to get your point across.

Blow your horn or wave your hands. If he suspects *anything* wrong, he will be grateful for the feedback and will pull over promptly. Most truckers now have two-way CB (citizen's band) radio units on board and they pretty much watch out for each other, but you can help if you spot something out of the ordinary.

As a matter of fact, individual CBers and the large CB organizations like ALERT and REACT do a lot of good in motorist assistance and early notification of accidents or danger. If you're involved in a road accident far from a telephone, look for a car or truck sporting a CB antenna. This will give you instant notification for fire or police assistance in your emergency. But flag down the first motorist, in any event, for notification or assistance.

Gasoline trucks today are still not made like the fuel trucks of the stricter airports, which self-seal if they rupture or begin to leak. They may be compartmentalized but, if you see something going drip, drip, drip out of one of these, you should get the driver's attention and get the hell out of there and call the fire department.

If the leakage from a tank truck or a railway tanker is substantial, you don't need to wait for some authority to order an evacuation. Get the hell away from there and tell everybody else to do the same. If the thing is leaking a gallon a minute or just flowing, go a city block, a football field away, or move *uphill* away from it. That thing could be pouring into sewers or storm drains downhill faster than you can walk. How far and how fast depends on what seems to you to be happening and how alarmed you are about it. If possible, make sure the driver knows about it, especially if it is parked and he is in a diner or something.

The difference between an explosion and a disaster is where the truck is. Out in the boonies or on an interstate highway it is one thing, but thousands of gallons spilling into a downtown storm drain could amount to a massive human catastrophe.

The same rule applies to ships in port, grounded airplanes, railway cars—and it isn't limited to gasoline leaking. A chlorine tanker overturned on the beltway a couple of years ago and a very large area had to be evacuated. There was life loss anyway, but it could have been a lot worse if there hadn't been immediate, widespread evacuation.

Truckers with dangerous loads should call police or fire people before entering any large city and take a prescribed route through or around it. Some cities will give them an escort. Most people are blissfully unaware of what's rolling by them—all the time.

On boats—even small pleasure boats—gasoline or gasoline vapor leaking into the bilge, especially during or just after fueling, is a serious potential for fire.

At the time of filling, the motor should be turned off and the battery disconnected; all power systems, generators, and motors—even air conditioners—should be shut down; obviously no open flames or smoking should be going on. That includes cooking anywhere on board or nearby. The bilge should be vented and carefully checked out before the motor is restarted. All spillage should be washed away.

The electrical system should be waterproofed, grounded, and sparkproof. This is all under Coast Guard jurisdiction and they really spell it out so that if you don't know and don't comply with these things, you may lose, or not gain, the right or license to operate a boat. And anybody on a boat is bound by these rules; it is the captain's responsibility to see that all passengers comply.

There should definitely be a CO_2 or DC fire extinguisher on board. If not legally required, it is nevertheless an absolute must.

Again, if you have an electrical fire, disconnect the current, and proceed with extinguishment.

And remember, gasoline floats in the bilge and on top of the surrounding water if there is enough spillage. If a boat goes up or burns, gasoline leaking from it can spread on the surface of water. So get completely clear.

A ship-to-shore radio is a handy thing to have for notification in any kind of distress. It could save your life.

Here, too, don't smoke in bed, or even in the cabin, if everything isn't safely vented.

CHAPTER 27

Eating Out

Although I have no figures to back it up, I assume that more people eat out than go to movies, theater, and church combined. I am going to further assume that they do it more often.

In 1972 (the most recent figure available), 21,000 fires occurred in restaurants. There were slightly under 250,000 restaurants in this country in 1972. Therefore, about one in every ten restaurants had a fire in 1972.

Were you in one of them?

Were you in one of them during a fire?

You're also probably thinking of Sloppy Sam's with the grease all over the walls and the flies in the window.

In 1972, the restaurant on the ninety-sixth floor of a large office building in Chicago caught fire. This is the highest high-rise fire to date. Conscientious preplanning saved it from catastrophe and there were no lives lost.

Twenty-five persons died in the 1967 penthouse restaurant fire in Montgomery, Alabama. This tenth-floor restaurant was of relatively fire-resistant construction: protected steel frame with brick and clay panel walls and a concrete roof. The second enclosed stairwell in the building did not extend to the penthouse level. The ceiling partitions and much of the furnishings and decorations were highly combustible. The penthouse had no sprinkler system. Two self-service elevators with nonfire-rated, self-closing doors ran from the basement to the penthouse. Six firemen were trapped in these elevators, which responded to a signal from the penthouse level. Both elevators stalled there, could not be moved, and their doors would not open.

It started as a small fire in the cloakroom, which no one warned customers about; no one asked them to leave. The fire quickly grew beyond the chef's attempt to extinguish it with a small fire extinguisher.

Then the fire suddenly flashed to the ceiling and then to the kitchen. Suddenly, the entire penthouse was in flames. There were about fifty people present when the fire started. Twenty-five of them died. If that doesn't scare enough hell out of you to make you think, I don't know what will.

Now for the good news. Nearly three quarters of restaurant fires started when these restaurants were completely vacant. Restaurants have a high incidence of incendiary fires; about one third of them are arson, followed by about a quarter caused by electrical irregularities, another quarter started by cooking and exhaust-duct fires. About 7 per cent are caused by smoking.

A third of them start in the kitchen (there's a lot of fuel and a lot of fire and a lot of heat in any restaurant kitchen), and about 15 per cent begin in the dining area.

What does all this have to do with you? I grew up feeling pretty secure in all restaurants. They looked pretty safe to me, really warm and comfortable with the little candle on each table, the frilly white tablecloths, the intimate darkness.

Now that I've seen a few of them after the fire and some of them during the fire, I'm more deeply comforted by the lighted EXIT sign over the door, and I really relax when I don't see a table in front of it.

Candles on a restaurant table are really dangerous. In some places, they're illegal, except for the hurricane or deep glass-enclosed lamp. They have flame in them, but at least they don't come into direct or close contact with the tablecloth or anything flammable. Still, spilling one of those things can be worse than upsetting the gravy.

Flaming foods served before your very eyes are foods on fire anyway you cut it. I once saw a guy in a restaurant trip and fling a flaming dish halfway across the room. It landed on the rug and he had the presence of mind, despite his embarrassment, to run across the room and kick it at the open fireplace. He missed the wicket and ended up stomping it to death. He was lucky; no one panicked. They were too busy trying not to laugh.

That laughter could have just as easily turned to sweat or tears if a tablecloth had caught on his pants leg or it had landed in a lacy lap. Brandy burns like crazy; it's approximately half alcohol.

You will remember I've cautioned against charcoal cooking inside almost as many times as I've mentioned gasoline as being the prototype no-no. You're probably wondering how the restaurants get away with their big charcoal fires barbecuing away so sizzlingly; it's because those are not charcoal briquettes. It usually is gas fired and made to look like the real thing.

Ordinances vary locally, but chafing dishes and such, if used at all, should be placed on a cart, never on the table. They should never be carried flaming from table to table. The food should be served only when the flame goes out. Christmas trees are pretty, too, and look at all the grief we get from them. Crepes suzettes, fondue pots, coffee warmers and the like are likewise dangerous.

There should be a seating-capacity notice posted in at least one conspicuous place in every restaurant. This notice limits by law the maximum occupancy of the restaurant. That doesn't necessarily mean people can't line up standing—inside from the cold—if they are orderly and they don't block the exits. As a patron, you have to be very careful in an overcrowded situation like this. It is important to know whether you can get out that exit if you have to; to know whether people, tables, or chairs, or anything else, are blocking any exit. I'd hate to have to run my family through the kitchen if the kitchen was on fire and the other exits were blocked by hysterical people. That's what happened in that Louisiana penthouse restaurant. A group of those people died in that kitchen. They never had a chance because they couldn't get out and they tried everything including running right to the fire. Panic is a terrible thing. So is fire.

What do you do when you walk into a restaurant and it just doesn't feel right? It's obviously overcrowded and you feel as if you're walking on people's plates just getting to your table, and you have to say three "scuse me's" just to get in your chair. Should you stay? Leave? Say something? Some stay (obviously). Some leave (inconspicuously). Some complain (and it ruins their evening).

I'm not thinking about a fireman or an inspector either. I'm thinking about you. You take your family out to dinner. Everybody has his heart set on this restaurant in particular. It's just so right for this particular evening. So you get there. It is a very popular restaurant and there are lots of other people there who also decided this was the place.

You walk in and sit down and you look up at the notice and it says 132 people are allowed to eat in this room. The waitresses are busy and service is a little slow, so you count how many chairs are occupied—137. You start thinking about that and you count again and you come up with 142 this time and you haven't had your first drink yet. Also out in the lobby there are another fifty to one hundred people waiting in a sort of fat crooked line. This is the exit; you came in that way and you don't see another way out without squinting. This would be your first-choice exit—maybe your only—if someone yelled fire or you smelled smoke. And you do smell smoke, but it's probably the duckling á l'orange the third table up. Isn't that nice? Looks good. Then all of a sudden it looks like a World War II newsreel flashed across your memory.

The waitress arrives and wants your drink order. Should it be two Shirley Temples, a Coke, and a couple of Molotov cocktails?

You think. You start thinking real hard and you break out in sweat.

"What's the matter, sir?"

"Dear, what is it?"

At this point you need a drink and think a toast to Coconut Grove would be appropriate so you . . . what?

What do you do?

Ask the waitress if there is a sprinkler system, whisper *fire,* ask for the manager, get sick, or have three drinks to catch up with the rest of these people.

Suppose you ask the waitress if there is another exit.

"The way you came in, sir."

You say, "*No, another one.*"

"Second door to your right up the same hall."

"*No,*" you say. "*A fire exit.*"

"Fire, sir. We don't have a fire. We never had a fire here in the eleven months I've been with them."

"*I'm glad you don't have a fire. I just wanted to know where the fire exits are in case you do have a fire.*" At this point, if the word "fire" is mentioned one more time, there could be a panic. Already, some people have stopped chewing to listen in.

"Over there, sir."

"*I beg your pardon, over where?*"

"Behind those maroon drapes, sir."

Please, don't point! "*I think I see it. Is it to the left of those stacked chairs?*"

"No, sir, to the right of the coffee urn, behind the temporary pastry table—right behind those drapes. You can't miss it. It's got a sign. The law, you know. What would you like to drink?"

"*Is that the only way out of here?*"

"No, sir, I told you the way you came in is the best way out at the minute. I'll get the manager. He'll explain it to you."

Your wife is embarrassed. Your children are embarrassed. You are embarrassed, but you are also getting good and angry.

It is important not to cause a panic by any further conversation with the word "fire" in it. You decide to intercept the manager before he comes to your table, so you tell your family to please follow you to the entrance and you'll meet them there.

Halfway up the waiting line you see a manager type. He doesn't look like the sort of person who could take your life and your family's lives so lightly.

"*Are you the manager?*"

"Yes. Are you the party who was asking about the exit?"

After all this discomfort, why not go all the way. If this hadn't happened, you might have just sat there and got indigestion or just slunk silently away. Now, it has all been made easy for you to follow through. You have nothing to lose, since you have already lost your appetite and your twinkle. You already have been embarrassed beyond words and now your anger is clear and you really know why you are angry.

"This place is overcrowded and your fire exits are blocked. You've got a candle going on every other table and the employees would go deaf, dumb, and blind if a fire broke out. I'm going to call the fire department at the nearest phone and I don't think I'll bring my family in here again after seeing what you think of your customers' lives."

"Nothing personal, sir."

"You're goddamned wrong, it *is* personal."

At this point your family joins you and you lead them to the outside door without another word. Some people in the line waiting to go in are looking at each other, but no one moves. They heard.

"But, sir, what do you want me to do? I can't ask my customers to leave."

Bars, lounges, night clubs, and discotheques present the same problems, sometimes worse with wall-to-wall people, a haze of smoke thick enough to trigger any smoke detector, or conceal a smoldering fire for quite a while. The other exit is through the kitchen which is closed this late at night. To keep out gate crashers, the other doors are locked for reasons of "security." Don't you know it's against the law for minors to drink here?

Some of the smoke is from candles on the tiny tables. The music could mask a grenade exploding. The revolving colored light is hypnotic. The red crepe paper on the other bulbs are scorched and freaky. The tablecloths are full of cigarette scars and the ashtrays are overflowing. The bartender dumps ashtrays from the bar into his all-purpose trash can. The booze is really flowing and everybody is having a good time. Looks like a real fun place. Nothing serious going on here. Everybody's having a ball.

So, all right already, it's only on Saturday nights. We're dead the rest of the week.

I have no available statistics on Saturday night fires—late, in a place like this.

Do you really need one?

I would strongly urge that you leave at once. (What are you really doing there anyway?)

Call the fire department, the police, the sheriff—anybody. You've just saved *your* life. Why stop there?

No Goody Two-shoes about it. If you saw a fire you'd report it and really feel great about maybe saving somebody's life. Well, you just saw one about to happen. Because there was so much noise, it was too loud to alert anyone. You just escaped and assembled.

Now, notify!

Reporting a fire that's about to happen is worth doing if you have the responsibility level to do it. Don't report it as a fire.

Just tell the fire people what's going on and how serious you think it is and tell them you'll hang around outside and wait for them.

I think that takes some guts.

And I think nobody will offer thanks.

EXIT
sheppard

CHAPTER 28

The Early-warning Detector

Three basic types of fire detectors are commonly available: heat detectors, smoke detectors, and flame detectors. The use of flame detectors has little or no practical application in the home. It is not my intention to sell you any of these special products, but buying that extra early warning is vital to your surviving your fire.

There are basically two effective types of heat detectors:

The rate-of-rise detector would be more reliable in rooms, such as attics, where normal temperatures on hot days could approach that of the alarm-activating temperature. The fixed-temperature heat detector is very much like the thermostat that controls your heating system. Some are activated by a fusible link that melts at a preset temperature in much the same way as a fuse in an electrical fuse box. Heat detectors provide early warning of hot fires in their immediate area, and only that. They do not respond in any way to smoke and would be of little use to signal smoke accumulated at any distance from the heat source.

It becomes clear that the smoke detector is more ideally suited generally for survival in the home, although it must be clearly understood that the ideal detector, perfectly sensitive to all conditions of fire, does not exist.

A smoke detector is designed to detect the presence of visible or invisible particles of combustion, and smoke in the surrounding atmosphere; it will signal an alarm when the level of these gases reaches a predetermined danger point.

These two types of smoke detectors are available as single-station smoke detectors. They are also available as complete systems, consisting of several units connected to a common alarm that rings, say, in the master bedroom or in all bedrooms, and to a general alarm connected directly to the fire department. If any single unit within the system senses fire or smoke, it will activate its entire alarm system. These systems are expensive, costing $400 and upward. Here, we are primarily concerned with the more affordable single-station smoke detectors,

There are basically two types of smoke detectors:

1. The optical or photoelectric smoke detector.
2. The ionization-chamber smoke detector.

Each type of the single-station smoke detector has a sensing chamber, an alarm-sounding device, and an electric stepdown transformer or battery source within the unified detector housing. These detectors are designed to be wall or ceiling mounted, on the surface or flush with it.

The photoelectric smoke detector utilizes the reflected- or scattered-light principle. Under normal conditions, the light shines directly across the chamber uninterrupted as it travels to the light trap on the opposite side of the unit. When smoke enters the unit, the light trap detects a variation of light received from the source, at a preset level, the alarm will then sound.

The ionization-chamber smoke detector contains a *harmless* radioactive source material that ionizes (magnetizes) the air within the sensing chamber. This ionization process causes a small electric current to flow. It is when smoke particles from a fire enter the unit that they cause a reduction in the flow of the current; that presence is sensed by electronic circuitry and, at a critical point, it triggers the alarm.

Within its sensitivity range the ion-chamber detector will also pick up on products of combustion other than those directly emanating from destructive fires. At first this might seem to be another nuisance, but think about it and remember that smoke and poisonous gases usually connected with fire kill more people than fire itself. So you may have to adjust the ion-chamber detector close to the range just above normal cigarette or tobacco smoke levels, or cooking or fireplace smoke, and experience an occasional inconvenience when someone burns the toast or chain smokes.

You can opt for a photoelectric device if you have an efficiency apartment or you can tune up your ion-chamber sensitivity range. They're not all alike; you will have to learn and love your detector, whatever it is, and know what to expect from its specific personality and adjust your response to its. It must operate within a reliability tolerance that cannot permit or provide a false security.

At present, the incandescent-light source used in some photoelectric units has a life span of three to five years. When the bulb fails, the detector will trigger a trouble light or signal and the bulb can be changed. Each of these detectors should be equipped with at least one spare bulb. This is no worse a problem than changing a battery in the ion-chamber type and the need is less frequent. Some newer photoelectric detectors on the market utilize a solid state L.E.D. light source which replaces the incandescent bulbs. With these units the bulb replacement problem is eliminated.

The ion-chamber detector uses a radioactive source material that will outlast any building in which it is installed. Since it uses very little electrical power, with little stress on its circuitry, it will last a long time with little maintenance.

All things considered, neither detector is better than the other under all fire conditions in the home. Because there are so many variables, so many possibilities and combinations of flammables and ignition sources in the home, it is impossible to accurately predict which type of fire you will have. *Either* type is good for residential early-warning fire detection purposes.

There is no question that all fire detection devices will continue to improve—and rapidly. It is not to your advantage to wait for perfection; your fire could happen first.

How do you choose your detector? I hope I've given you more to go on than you had before.

There is much attractive and misleading advertising in magazines and newspapers that promises safety through this device or that device and is hammered in by a hard sales pitch based on scare tactics. This has always existed in the marketplace.

Clever, unethical salesmen have always made high profits by pushing overpriced or inappropriate devices that may fall short of the claims made for them. In some cases, the product may be worse than useless, it may be downright dangerous and, at the very least, give the buyer the dangerously false sense that he has fulfilled an essential life-saving requirement of his family.

It is a shame that it tends to have its greatest effect when the commodity is one of life-or-death importance. If you suspect that any salesman is putting on the pressure, your best bet is to assume a no-buying attitude and keep looking, listening, and reading. The better informed you are, the better choice you will make in terms of product reliability, dollar value, and home security.

If you are shopping for a complete system, I would suggest that you get prices from at least two companies that will install the same basic type of protection. You shoull consider the reliability of the installer as well as the product he proposes to install. Remember, your life may well depend upon his ability to install the product properly and provide a competent maintenance program.

You don't need an installer for the single-station detector, unless you buy a model that must be wired in by a licensed electrician. You can install the plug-in model and the battery-powered detectors yourself. Do not buy any device that is not backed up by evidence of reputable testing by qualified agencies. Don't take anybody's word unless they have the proof in black and white.

There are several things you can do to assure yourself that you are making the right choice. Inform yourself, read, check out the major consumer reports, and ask a lot of questions. Check the company out with the Better Business Bureau. Be satisfied.

A major guideline to follow in determining the reliability of your early-warning detector is to be assured that the product has received the approval of a recognized testing laboratory. This seal of approval should be affixed to the product. The Underwriter's Laboratories Inc., Underwriter's Laboratories of Canada, Factory Mutual System, or CSA can be considered to be reliable.

In my opinion, it is imperative that the product carry one or more of the laboratory approval seals previously mentioned. The seals may simply state approved by UL, ULC, FM, or CSA in some type of insignia. Quite obviously, this takes a great deal of guesswork out of your selection and gives you a most reliable assurance that the unit will operate as designed for in the time of your emergency.

In nearly all situations, the smoke detector should be located within thirty feet of each bedroom, located so that smoke from any fire that originates outside the bedroom area must pass over the smoke detector before entering the bedrooms. This location should detect a fire before the path of exit from sleeping areas becomes impassable due to the products of fire and smoke. It should assure sleeping occupants the earliest possible warning of fire danger.

In each bedroom where anyone smokes at night, there should be an additional smoke detector. I think you can easily picture your problems in having the smoke between you and the detector.

Many new building and fire codes now legally demand the installation of fire early-warning detectors in all new construction and certain types of older buildings. They generally specify the mandatory use of the home's "AC primary power source" to operate the detector, but permit the use of battery-operated units where connection to an AC source is not practicable. I completely disagree with this. I think this is backward and I know it is dangerous.

Now, here's the bad news:

Fire can knock out house current by burning out and short-circuiting electric wiring. An external breakdown or failure of power supply to the building—such as power interruption from the utility company for any reason or a tree across the power lines in a wind or a storm—could render this type of device useless, possibly when you most need it.

I think these codes are absolutely correct in their basic assumption, but dead wrong, dangerously wrong in their applications, for the reason that power failure leaves you worse off than you were with no smoke detector. It leaves you more deeply vulnerable to your fire because of the sense of security, your deep reliance on your detector to save your life. This house-current factor becomes the weakest link because it exposes you to the high risk of *false security*.

In actual practice, battery operation has two strong advantages and one disadvantage over house-current-power operation.

The battery-powered device cannot be interrupted by any power failure inside or outside the home. It is so completely portable it can be easily mounted in the best possible location to detect a fire—and that is so important.

The major disadvantage is that these batteries will need periodic replacement, generally once a year. Space-age improvement in batteries makes this model very effective. There are basically two types of acceptable batteries: alkaline and mercury.

All battery-powered detectors are required by law to emit "a distinctive audible trouble signal" that will signal at least once each minute for seven consecutive days "before the battery is incapable of operating the device(s) for alarm purposes." This trouble signal will sound out as the batteries approach the end of their useful life for whatever reason (aging, terminal corrosion). In actual fact, it is no more possible to ignore it than it would be to ignore a baby screaming to have its diaper changed. This trouble signal must be conspicuously different from the detector's fire alarm, which in turn is required to sound continuously for at least four minutes, or as long as sufficient smoke is present to trigger it.

CHAPTER 29

Now It's Up to You

We had a nursing-home fire recently that was outright scary!

As we know, the people who live in nursing homes are much more vulnerable than the ordinary citizen and they require more assistance to evacuate such a fire and even to survive the evacuation. And we also know that the result of a major fire in a place like this can be tragic, disastrous. On January 9, 1970, an Ohio fire killed thirty-one of forty-six patients present in a nursing home. A Kentucky nursing-home fire killed ten and injured fifty in January 1971. In September 1971, a nursing-home fire in Pennsylvania killed all thirteen patients; they had all been so heavily sedated at bedtime that they could not even be alerted in time to escape.

The story could be much the same in any establishment of this type, at any time.

The fire department has tried educational programs. Many other agencies offer information and advice, but so much of it is so piecemeal and incomplete that each piece is only a fragment. The whole answer is yet to be assembled at one time, in one place.

The fires continue, while everybody looks far and wide for the answer lying right under their noses.

I had a nursing-home fire and I can tell you that going into a building like that and seeing 112 people being evacuated, seeing smoke coming out of the fourth floor in that heavy volume was like a nightmare you thought could never be real.

Fortunately, in this fire this time, there was a smoke detector on the fourth floor of the building where the fire started, in a linen closet. It automatically dialed the fire department. We arrived at that fire within sixty-five seconds. We entered the building before the people inside even knew they had a fire.

You and I both know any fire department has the responsibility and capacity to respond to any fire, anywhere, anytime. But beyond our responsibility to go there and put it out, to rescue survivors, what else are we supposed to do?

I believe there is a twofold obligation involved here. I must emphasize one last time that before we can respond to a fire, we've got to know there is a fire, and where. Before we can help you, we've got to know about your fire. You are going to have to start doing something to help yourself.

Most of the deaths have occurred at night. If you're asleep and you don't get some warning and you don't get a chance to move before the smoke and the fire rise to where you are, you're not going to get out; you're going to become a statistic.

If you do wake up with the early warning or the smoke and you haven't preplanned, if you don't have some idea of what you're going to do, you're going to stick your head up into hot smoke and you're going to panic. And everybody around you is going to panic. And the next thing you know, you're going to do something foolish like jump out of a fourth-floor window headfirst, throw a child out of a window, or open a door and go running into the blaze; but if you had preplanned on the basis of an early warning you would have responded instinctively to your preplan and given yourself your chance to come through your fire alive and unhurt. And you would have had the once-in-a-lifetime joy of meeting your entire family outside the fire, instead of a permanent grief.

We haven't mentioned to any great extent the hundreds of thousands of persons who have ended up seriously burned, tortured, disfigured. Some die weeks or months later in a burn or shock trauma center in some hospital, which can never even temporarily substitute for a home. I'm not going to show you pictures or tell you how expensive this is in every possible way. We still know far too little about burn treatment, but we know it calls for treatment of the entire person and not just his surface. The life that is physically saved is far too often "ruined" if the pain and disfigurement are too great, for too long. The human spirit can survive only so much bombardment before it becomes crushed. Even the spirit of the saviors in these burn centers suffer much abuse and sometimes can take no more. In many burn-care facilities, there is a 100 per cent turnover in nursing staff every six months. Doesn't that tell you all you need to know for you to do everything in your power to avoid this fate for yourself, and for your family?

And guilt. Do you have any idea how many tortured lifetimes are spent in anguish over a small but deadly mistake in judgment made at that very moment when one is least equipped to judge clearly and with reason—awaking in horror to a fire one never thought could happen—one for which one never prepared?

It strikes me now that we're getting into a whole new kind of situation across the country. As you know, the number of single-unit dwellings has diminished over the past years and seems to be phasing out in terms of new construction. The trend is now to condominiums, apartment houses, multiple-family rowhouses, and the like. This trend has completely reversed itself over the past ten or fifteen years.

Does that not now present a new kind of need for a different kind of fire protection?

I think it shows two needs. From the fire-protection standpoint, it definitely gives a new set of problems on how best to attack the fire once we get there and how to fight the fire. And also how we can more effectively engage the community to prevent the fire, and—equally important—how we can help them to train themselves to survive the fire and even in some cases to contain it until we arrive.

It changes your responsibilities around in a way you have never been taught to accept. Perhaps you have never even thought about *your* responsibility to yourself and your family in this context before, having been more or less satisfied to leave it all up to God and the fire department and relying, without thinking much about it, on luck or faith that it will all work out OK when the chips are down.

Your personal survival is completely dependent on the level to which you have accepted, and can use, your personal responsibility, regardless of any exterior factor.

And, if you move into an apartment-house complex, you now have a responsibility in your evacuation plan not only to alert your family but, as you make your personal escape, to signal a general alert any way you can without sacrificing the success of your own escape. If there is time to do it, it would be mighty thoughtful if you banged on some doors and yelled "*Fire*" on your way out. You're in a community arrangement now, and you can do unto some others what you'd be damned thankful they would do unto you if they have the first advantage of earliest warning. Slam down on all those buzzers as you pass the mailboxes on your way out.

That's the coming thing now! If you've got a fire, everybody in the building has got a fire. If someone else has got a fire, it's your fire, too.

But, now for the good news: If somebody else has got a detector, you've also got an early warning if the guy isn't a selfish rat. Likewise, you have a full responsibility to others in the building to share your early warning at the same time, or immediately *after*, your family alert has been sounded and responded to by instant evacuation.

And further, you and every other tenant in your building are due, are owed, a realistic briefing, by the owner or landlord, of the building prefire plan—escape routes, equipment, precautions, detectors, sprinklers, whatever is available at your point of occupancy and whatever is added or changed as time goes on. In moving in, your very first job is to establish your personal prefire and evacuation plan consistent with the available physical condition.

In any rented home, your landlord owes you this as surely as you owe him rent, for you are not just renting space; you are renting living space, and you are getting cheated if there is fire or death built into it.

There is a problem of owner apathy and that is going to have to change—right now, not maybe later. If the owner doesn't seem to realize what that rent money is for, that you, his tenants, are living people deserving as much protection as the flammable paper rent-receipt books he stores in a fireproof safe, then you are going to have to let him know you are alive by getting his attention. You are also going to have to do what you must do in any situation; you are going to have to take the responsibility to save yourselves.

Part of this is to notify the fire department of any violation of your safety if the owner is unresponsive or unapproachable.

The quicker the fire department gets there, the quicker they're going to put out the fire to save the property. Therefore, I would certainly think it would be to the owner's best advantage, his realistic self-interest, to encourage things like fire wardens, fire drills, smoke detectors, protection plans, and fire extinguishers. Nevertheless, so far it is extremely difficult to get owners to become involved. They don't live on the premises; they don't want to become involved. Something, however, seems to be working on the old apathy/indifference syndrome, that dark age of the spirit where people are not very responsible, even unto themselves.

"Fire" is a word that can strike terror in the heart of anyone who faces it personally. It is an unforgettable one-to-one relationship; if not terminal, it can furnish the driving force, the energy, and the mind-connections to increase one's chances in any future encounter.

However, the best teacher of fire prevention and survival is Mr. Fire himself. But he's a little selfish about graduating his students. He'd keep them after school forever if he could.

It is the purpose of this report to establish a valid substitute-teacher relationship, one to one, with you, the reader, in such a way that you can overcome your greatest handicap: the deep-seated and irrational belief that it won't happen to you. I find that this attitude of *sure, it happens to the other guy, but I won't let it happen to me* is common.

Now's your chance.

I've just about had my say. Within a very few pages, you will be on your own, which is where you always were and where you always will be. Now which do you prefer as tutor: Mr. Fire or the Fire Commissioner?

Although the public tide of apathy seems to be turning, that tide is still high. As long as apathy is high, fire will kill the apathetic and their more responsible neighbors along with them.

I think we are harvesting a crop of fire-prevention literature that has bounced off and deadened the public interest. This is the fruit we are bearing for trying to con the people into believing that all we have to do to put out fires is not to have them in the first place.

Everybody knows that we don't store a bunch of oily rags near the furnace. Everybody knows not to dump the ashtray in the wastebasket.

Everybody dumps these stick-figure little pamphlets in the wastebasket and enough of them go right on and continue to dump the ashtray right in on top of that. These pamplets give you at best a kind of halfway explanation and in many cases insult the intelligence or the sensibilities of the very person they're really trying to reach—the responsible person in the household.

It's time now. . . . This is a new world now. . . . People want to know why.

If you don't want them to get on an elevator when there's a fire, you've got to tell them why you don't want them to get on that elevator.

What happens with an elevator in a high rise is very simple: The elevator goes to the floor where the fire is, there's a draft in the airshaft where the elevator moves up and down, and then it very nicely opens into the blazing hallway, and you might just as well be in a sardine can inside your oven because there's no place to go.

That's why we don't want you in an elevator in a high-rise fire. But unless we explain both sides of it, instead of saying, "Hey, we want the elevators for the firemen to ride up in," the people are going to say the heck with that. "We want them to ride down in. It's our apartment house; we're paying the rent."

It's a matter of education, training, and common sense. People have been bombarded with answers that had no real connections to the questions or to each other.

There should indeed be a voluntary or mandatory retraining in the homes, the schools, the factories, to reappraise and reassemble what we already know and reapply it to where we really are.

I think it generally has to be one to one, personal, direct, and meaningful. Show and tell the people that if they take on the personal responsibility for the obvious, they stand a much better chance of surviving the fire that is statistically destined for each family, each generation. Sometime in your life you're going to be brushed by fire, grazed or totaled.

And if you know and plan ahead enough to realize that you can do something about it beforehand, you'll probably still be around afterhand.

Wherever you live, there are going to be basic inherent problems of design, construction, and materials, and you're going to be living with all kinds of solutions to these problems, from "very good" to "fire trap" or "booby trap." The age of the problem really has less bearing on your survival than certain design considerations. The open stairwell, from the basement to the roof, is a primary escape route that is going to be absolutely useless once that fire gets underway, within a matter of seconds or minutes. That's where the smoke and fire are going to be. You'd better damned sure have another way out—or two or three.

Many houses, like rowhouses, have windows and doors at the front and rear only.

If you're living next door, you'd better be concerned about the horizontal open "stairwell" effect; that fire can transmit in any direction once it dead-ends in its primary escape route. It will go up as fast as it can and then do what you should be doing, go directly toward the best alternate escape route. In the case of the fire, it may break through into the next attic or attics if there's not a fire wall between each house or unit. Having raced over next door, it can just as quickly start to work its way back down again in there.

You had better be well on your way out *your* alternate escape route—window, fire escape, whatever—for now you've got double trouble to say the least. Not just one house, but maybe the whole block or section is on fire.

In the case of "fire traps," inspection and condemnation are only as good as the codes and ordinances behind them. Any community with inhabited fire traps has the obligation to move people out or move correctional repairs in.

I fully believe that the owner or superintendent of any building has a great responsibility. I feel that any misrepresentation or deceit on his part about the safety of his building should be criminal; if he says he has an internal-alarm system or a sprinkler system or a fire escape, by God these things had better work, had better be what he says they are, or he should be in a whole lot of trouble.

And today, here as elsewhere, with the economy and unemployment as it is, somebody rents a house to a person and let's say the house has three or four bedrooms in it. Maybe a couple of years ago, you would have found five, six, seven people living here.

Now, you find not only the initial core family living here but some relatives and a couple of friends, who are on hard times, living here as well, all in a nice family atmosphere, except that when the fire hits, instead of having six or seven people to worry about getting out, you have thirteen, fourteen, fifteen people without a plan at all, running into each other instead of out. Not all of them sleep in bedrooms; some of them sleep on the couch in the living room or something like that. All that smoking in all those beds, no fire plan. Apathy, indifference, then fire, followed by confusion, panic, and death.

We have noticed a strange new risk recently, an increasing number of arsons in occupied buildings, which I will call an assault with a deadly weapon—fire. If they connect, that's murder; if the intended victim lucks out, that's attempted murder. It's as simple as that and the penalties must be that direct.

Arson is fiendish; it is the worst kind of murder. This isn't one-to-one murder; this is potential mass murder. The agony of a fire and its consequences for perhaps a whole family make an individual getting it between the eyes with a .45 look like a Sunday school picnic.

It's a terrible way to go, and incomplete survival can be worse than outright death.

Many of the restaurant and church fires are arson. And we firmly believe that more than half of all our fires of unknown or mysterious origin are arson. We can't even find the exact number of deaths caused by arson.

I have absolutely no sympathy for the arsonist; he is a madman and must be removed to where he can never light another match. We've got to protect ourselves against him, and prevent, pursue, capture, and prosecute this insane criminal to the fullest limits of our laws.

This is one reason why fire prevention alone cannot, I repeat cannot, assure you the best odds on your fire safety. You could have the cleanest house on the block. All it takes is one crooked nut and one stray match—and there you go. Protect yourself.

You see the average fireman and you talk to him in the firehouse and he's a tough, blustery guy, and he's going to go in there and he's going to show you nothing bothers him and he's going to go in there and do his job. And then you meet the guy the next time up at a fire scene and he's just come out with a child or an adult in his hands who didn't make it. I've seen some of the toughest 250-pounders sitting there on the steps with the soot in their eyes, their nose, and their mouth and somebody died on them. And let me tell you, that is personal, that is a loss for every fire fighter whether he was on that fire or he wasn't on that fire. He staked his life to rush there and, believe me, he takes it personally.

So, you are not going to bat a thousand, one way or the other, in a situation like this anywhere in the country. But our heart and our thanks certainly go out to all the fire fighters for the splendid job that they have done and will continue to do.

They can do that job better only if they can get there sooner. Someone has got to let us know. We have to know about it before we can go and put it out and maybe save your life. Any early-warning device can give that precious gift of time, that edge of minutes or even seconds against your fire's acceleration.

When you and your family are safe and together, quickly lend that gift to us. *Notify* and trust that we will use that precious time to do for you what you cannot do for yourself, to save what you have left behind when you so wisely chose to save yourselves.

It's no fun coming down the street in ten tons of apparatus and worrying whether or not the car in front of us, or the pedestrian, is going to move out of our way.

Every time you hear that apparatus going, whether it's a false alarm or a real alarm, we think we're going to a fire. Get out of our way and stay out of our way until we're clear of you. Don't stand between us and that fire.

Index

A

ABC extinguisher, 142, 145
Aerosol boat horn, 53
Aerosol bombs, 133
Air conditioner, 34, 35, 125
  overheating of, 36
Alcohol, 35, 120
  flash point of, 31
Alertness, 143, 147, 152–74
  office building, 154–61
  rules for, 152
  survival attitude for, 152–54
Animals and pets, 95–98
Appliances, electric, 34–35, 116, 117, 118, 120, 122, 125
  disconnecting, 123
Arson, 113, 134–37, 169, 194–95
  paid fire department investigation of, 137
  types of, 134–36
Ashtrays, emptying, 113, 192
Asphyxiation, 13, 27, 41
Assembly point, 44–45, 159
  purpose of, 95
Automatic Reporting Telephone (ART), 47
Automobile accidents, 171–72

B

Barbecues, 130
Bar fires, 179–80
Basement, 51, 133
  flammable items in, 120
Basement windows, 40, 51
Bathroom, electrical wiring in, 126
Battalion chief, 91, 94
  concerns of, 82–85
Battery-powered detectors, 186–87
Beds, 23, 27, 67
  mattress fires, 100–2, 112–13, 120
  smoking in, 36
Bicarbonate of soda (baking soda), 118, 145
Bingo halls, 169
Bottled gas, 34
Breathing carbon monoxide, 27
British Airways Terminal Building, 76
Brush cleaners, 120
Building fire codes, 78, 79–81, 84, 186
  landlord and, 80
  tenant association plan, 80
Burns, degrees of, 42

C

Candles, 128, 129, 130, 132, 180
  in churches, 170, 171
  restaurant table, 176
Carbon dioxide, oxygen and, 40–41
Carbon monoxide (CO), 16, 26–27, 40–41, 119
  charcoal as, 36
  as poisoning, 114–15
Ceiling fire, 143
Charcoal, 36, 119, 128, 176–77
Children, 16, 24, 37, 38, 39, 87
  on ladders, 66
  lowering from a window, 66–67
  lower-level evacuation of, 55, 56
  matches and, 103–10
    babysitters, 106–10

fire toll, 103
parents that smoke, 104, 105, 106
"playfulness," 103–5
teaching the child, 104–5
where they hide, 21–22
Christmas trees and decorations, 126, 132, 177
Church fires, 153, 169–71, 195
Cigarette smoking, 33, 45, 104, 105, 111–15
in the bedroom, 36, 101
of the elderly and handicapped, 111–12
and emptying ashtrays, 113, 192
falling asleep and, 111
fire toll, 111
mattress fires, 112–13
Cleaning fluid, 35, 119
flash point of, 31
Closed door, functions of, 14–16
Closet, 23, 27, 67
Clothing on fire, 130, 140
what to do, 145–46
Coughing, 18
Crowd control (at exhibitions), 169

D

Deaths by fire, 1, 2, 13, 19, 84
major cause of, 13
number of per year (U.S.), 1, 2
Department stores, 169
Discotheque fires, 179–80
Doors, 22, 32, 37, 42, 71, 74, 88–89, 150
cords under, 125–26
at the ground level, 51–52
kinds of, 24
in motel rooms, 59
opening (and when not to open), 14–16, 24–25
stairwell, 78, 79
stuffing towel under, 40
Drafts, 25, 76, 86, 87
Dry chemical (DC) extinguisher, 142–43, 145, 171, 174
Dumpster fires, 129

E

Early-warning detectors, 8, 11, 13, 18, 182–87
battery-operated, 186–87
guidelines for buying, 185
location of, 185–86
types of, 182–85
Electrical hazards, 122–26
repairs and maintenance, 123, 124
types of, 125–26
wiring installation, 122–25
short circuit, 123
Electric blankets, 125
Electric hot plates, 120
Electric iron, 124, 125
Electricity:
circuit overloads, 35–36
disconnecting, 142
and gasoline vapors, 32–33
ground-level wires, 51
short-circuited, 186
Electric ranges, 125
Electrocution, 143
Elevators:
in a high rise, 192–93
when not to use, 74–76
Empire State Building, 76
Evacuation. *See* Fire escape plan
Extension cords, 122, 125–26

F

Factory Mutual System, 185
False alarms, 19, 129
Fire alarm boxes, 45, 47, 148
in multifamily dwellings, 69–70
Fire and fires:
electrical hazard, 122–26
fascination with, 12
fighting, 138–96
by alertness, 152–74
early warning systems, 182–87
extinguishment, 141–46
or fleeing, 138–40
getting involved (as a passer-by), 147–51
personal survival and, 188–96
when eating out, 175–81

heating and cooking, 116–21
how they start, 30–36
combination factor, 30
electrical appliances, 34–35
electrical overloads, 35–36
flammable vapor, 30–35
as a killer, 13–17, 23, 84
matches and, 103–15
shouting "fire," 19, 88, 148, 165
sparks and, 127–33
that return, 100–2
types of, 142–43
Fire burglar alarm combination, 61–62
Fire chief. *See* Battalion chief
Fire codes. *See* Building fire codes
Fire Department:
getting help from, 47–49
responsibility of, 188–89
Fire drills, 19
Fire escape plan:
basics of, 4
jumping and, 61–81
when not to jump, 69–81
lower level (first floor), 50–60
preplanning, 21–99
evacuation, 50–81
getting help, 47–49
panic and, 86–92
rescue, 93–99
typical home situation, 21–29
purpose of, 1–6
reason for, 7–12
responsibility for, 3–4, 17, 18–20, 23, 37–38, 43–44, 57, 71, 95, 153, 190
Fire escapes (exterior), 61, 69, 87, 162
Fire extinguishers, 26, 138, 141–46
best recommended, 145
how to use, 143–45
rules for, 141–42
types of, 142–43
Firemen:
battalion chief, 82–85, 91, 94
rule of rescue, 68
salvage responsibility, 98
Fireplaces, 30, 104, 127–28, 130, 132
combustibles near, 127–28
flue, 127
as a killer, 36
screen, 128
Fire prevention literature, 2, 7–8, 11
Fire sparks, types of, 127–33
Fireworks, 128
First-degree burns, 42
First-floor dwelling. *See* Lower level evacuation
Flame detectors, 182
Flashlights, 41–42, 132
Flash point (or fire point), 31
Floor wardens, responsibility of, 157-58
Fondue sets, 128
Fuel storage, 57
Fuse box, pennies in, 123

G

Gases, 1, 16, 42, 140, 155
inhalation of, 13–14
as a killer, 13–14, 23
Gas leaks, 116, 119, 121
Gasoline, 35, 104, 105, 119, 128, 132, 171
camping stoves, 120
at dynamite status, 34
flammable vapors of, 32–34
flash point of, 31
fumes from, 132
Gasoline tankers, 172–74
bilge leakage (in port), 174
on the road, 172–73
Getting involved (as a passerby), 147–51
Grass fires, 129

H

Hair lacquer spray, 128, 133
Heat:
pressure built up by, 14, 16–17, 33
travel of, 28
Heat detectors, 18, 47–48
types of, 182
Heaters, 125
Heating and cooking, 116–21

basement or cellar, 120–21
kitchen stove, 116–20
do's and don'ts of, 119–20
wood stoves, 127
Heating pads, 125
High-rise building fires, 157, 175
elevator, 92–93
roof and, 77
See also Multifamily dwelling evacuation
Household materials, flash point of, 31

I

Individualism, 164–65
Inspections, fire-prevention, 84–85, 157
Insurance companies, 81
Ionization-chamber smoke detector, 183–84

J

Jumping to escape, 37, 61–81
alternatives to, 63–68
chances of survival, 69
and landing pile, 66
and rescue, 93–94
when not to jump, 69–81
See also Windows

K

Kerosene, 34, 120, 128
Kitchen fires, 57–58
Kitchen stove, 116–20, 127, 130, 133
pot of food on fire, 118
wood- or coal-burning ranges, 119–20

L

Lacquer, 35, 120
Lacquer thinners, 120
Ladders, 51, 63, 139–40
climbing, 64
fire-truck, 84–85
Lamps and light fixtures, 125
Leaves, burning, 128–29
Lounge fires, 179–80
Lower level (first-floor) evacuation, 50–60
of children, 55, 56
distance and time factor, 55–57
of the elderly and invalid, 53–55
factors common to, 50–51
hazard in, 51
jumping, 63
mobile homes, 57–58
motel room, 59–60
number of exits and, 52
recreational trailers, 58–59
types of dwellings, 50

M

Manual of Policy and Procedure (Fire Department), 82
Matches, 103–15, 116–17
children and, 103–10
babysitters, 106–10
fire toll, 103
parents that smoke, 104, 105, 106
"playfulness," 103–5
teaching the child, 104–5
cigarette smoking and, 111–15
ashtrays, emptying, 113, 192
elderly and handicapped persons, 111–12
falling asleep, 111
fire toll, 111
Mattress fires, 100–2, 112–13, 120
Mobile homes, evacuating, 57–58
Motel room, evacuating, 57–58
Movie theater fires, 153, 165–68
Multifamily dwelling evacuation, 61–81
alarm system and, 69–70
elevators and, 74–76, 92–93
fire codes and, 78–81
responsibility for, 71
stairwells and, 70–79
when not to jump, 69–81
when trapped (what to do), 73–74
window exit, 61–68

N

Night club fires, 179–80
Nursing-home fires, 188

O

Office building fires, 154–61
  assembly point, 159
  floor warden duties, 155–58
  how they start, 154–55
  what to do, 160–61
  on your floor, 160
Oil-burner backdraft, 120–21
Oily rags, 192
  ignition temperature of, 33–34
Overload, electrical, 35–36
Oxygen, 16, 25, 27, 118, 141
  and carbon dioxide, 40–41
  decrease, 13
  with vapor, 31–32
  from windows, 41

P

Paint, 34, 120
  flash point of, 31
Paint remover, 132
Paint thinner, 34, 35, 120
  flash point of, 31
Panic, 83, 86–92
Paper, ignition temperature of, 33
Pets, 95–96
Photoelectric detectors, 183–84
Pilot lights, 34, 51, 57, 87, 133
  when on vacation, 121
Plate-glass windows, 62
Plexiglas, 40
Police, duty of, 148
Portable heaters, 126
Profit type arson, 134–35
Propane storage tank, 133

R

Radios, 126
Rate-of-rise detector, 182
Recreational trailers, evacuating, 58–59
Refrigerators, 125
  motor spark, 119
Reporting a fire, 47–49
  alarm box, 47
  by detection devices, 47–48
  telephone, and what to say, 48–49
Rescue, 93–99, 140
  methods of, 93–94
  of people first, 95–99
Restaurant fires, 153, 175–81, 195
  number of (in 1972), 175
Roofs, 28, 38, 63, 69
  of adjacent houses, 67
  as an exit, 26
  in high-rise buildings, 77
Roof ventilation, 27
Rope ladders, 63–64, 66
Rugs, cords under, 125–26

S

Safety ladder, 26
  disadvantages of, 63–64
School fires, 161–64
  how they start, 161
Screen, fireplace, 128
Screens, window, 39–40
Second-degree burns, 42
Sheets, tying together (to escape), 64–66
Shock, 23, 140
Single-station smoke detector, 183
Smoke, 22, 42, 87, 88–89, 139, 143, 150
  buildup of, 16–17, 27
  inhalation of, 13
  as a killer, 13, 23, 26–27, 101
  travel of, 28, 76–77
Smoke detectors, 2, 11, 18, 29, 47–48, 57, 81
  types of, 182–85
Smoke explosion, 25
Smoking. *See* Cigarette smoking
Smoldering mattresses, 100–2, 112–13, 120
Sprinkler systems, 69
Stairwells, 70–79, 86, 87, 96, 114, 162, 193

AC/DC emergency lighting system, 78
air pressure in, 78
door and, 78, 79
interior standpipe connections, 79
Storm windows, 39
Sulfur candles, 129

T

Telephone, for reporting a fire, 48–49
Television sets, 126
Tent fires, 168–69
Third-degree burns, 42
Titanic (liner), 76
Torches, makeshift, 130–32
Towering Inferno, The (motion picture), 157
Trash, burning, 128–29
Truman, Harry, 157

U

Underwriters' Laboratories of Canada, 185
Underwriters' Laboratory (UL), 123, 126, 132, 185
U. S. Coast Guard, 174

V

Valuables and possessions, 98–99
Vapors, flammable, 30–35
combination factor in, 30
fireplaces and, 30, 36
flash point of, 31
gasoline, 32–34
household materials, 31
oily rags, 33–34
with oxygen, 31–32
paper and cigarettes, 33
Varnish, 35, 120
Varnoline, 120
Ventilation, 28, 36, 61, 71, 143
Volunteer fire departments, 139

W

Wall paneling, do-it-yourself, 124
Washers and dryers, 125
Wastebaskets, emptying ashes in, 113, 192
Water, 139, 140, 141, 142, 143
Water heaters, 120, 121
Wet towels, using, 40
Window fans, overheating of, 36
Windows, 22, 25, 27, 28, 42, 51, 86
bars on, 40, 62
basement, 40, 51
breaking glass on, 38–39
cords under, 125–26
evacuation, 37–40, 61–68
of children, 66–67
rope ladders, 63–64, 66
safety ladder, 63–64
tying sheets together, 64–66
as an exit, 25–26
jumping from, 61–63
important factor in, 66–67
in motel rooms, 59
oxygen from, 41
plate-glass, 62
screens on, 39–40
teaching children to unlock and open, 38
Wiring installation, 122–23
World Trade Center Building, 76